Functional Analysis

Pure and Applied Mathematics

A Series of Monographs and Textbooks

Editors **Samuel Eilenberg and Hyman Bass**

Columbia University, New York

RECENT TITLES

MORRIS HIRSCH AND STEPHEN SMALE. Differential Equations, Dynamical Systems, and Linear Algebra

WILHELM MAGNUS. Noneuclidean Tesselations and Their Groups

FRANÇOIS TREVES. Basic Linear Partial Differential Equations

WILLIAM M. BOOTHBY. An Introduction to Differentiable Manifolds and Riemannian Geometry

BRAYTON GRAY. Homotopy Theory: An Introduction to Algebraic Topology

ROBERT A. ADAMS. Sobolev Spaces

JOHN J. BENEDETTO. Spectral Synthesis

D. V. WIDDER. The Heat Equation

IRVING EZRA SEGAL. Mathematical Cosmology and Extragalactic Astronomy

J. DIEUDONNÉ. Treatise on Analysis: Volume II, enlarged and corrected printing; Volume IV; Volume V; Volume VI

WERNER GREUB, STEPHEN HALPERIN, AND RAY VANSTONE. Connections, Curvature, and Cohomology: Volume III, Cohomology of Principal Bundles and Homogeneous Spaces

I. MARTIN ISAACS. Character Theory of Finite Groups

JAMES R. BROWN. Ergodic Theory and Topological Dynamics

C. TRUESDELL. A First Course in Rational Continuum Mechanics: Volume 1, General Concepts

GEORGE GRATZER. General Lattice Theory

K. D. STROYAN AND W. A. J. LUXEMBURG. Introduction to the Theory of Infinitesimals

B. M. PUTTASWAMAIAH AND JOHN D. DIXON. Modular Representations of Finite Groups

MELVYN BERGER. Nonlinearity and Functional Analysis: Lectures on Nonlinear Problems in Mathematical Analysis

CHARALAMBOS D. ALIPRANTIS AND OWEN BURKINSHAW. Locally Solid Riesz Spaces

JAN MIKUSINSKI. The Bochner Integral

THOMAS JECH. Set Theory

CARL L. DEVITO. Functional Analysis

MICHIEL HAZEWINKEL. Formal Groups and Applications

In preparation

SIGURDUR HELGASON. Differential Geometry, Lie Groups, and Symmetric Spaces

Functional Analysis

CARL L. DeVITO
DEPARTMENT OF MATHEMATICS
UNIVERSITY OF ARIZONA
TUCSON, ARIZONA

1978

ACADEMIC PRESS   New York   San Francisco   London

A Subsidiary of Harcourt Brace Jovanovich, Publishers

ACADEMIC PRESS, INC.
111 Fifth Avenue, New York, New York 10003

United Kingdom Edition published by
ACADEMIC PRESS, INC. (LONDON) LTD.
24/28 Oval Road, London NW1 7DX

Library of Congress Cataloging in Publication Data

DeVito, Carl L
    Functional analysis.

    (Pure and applied mathematics, a series of
monographs and textbooks ;)
    Bibliography: p.
    1. Functional analysis. I. Title. II. Series.
QA3.P8   [QA320]   515'.7   78–3333
ISBN 0–12–213250–5

Contents

Chapter 3 Linear Functionals

Chapter 4 The Weak Topology

Chapter 5 More about Weak Topologies

Chapter 6 Applications to Analysis

Chapter 7 The Theory of Distributions

Preface

Functional analysis, partly because of its many applications, has become a very popular mathematical discipline. My own lectures on the subject have been attended by applied mathematicians, probabilists, classical and numerical analysts, and even an algebraic topologist. This book grew out of my attempts to present the material in a way that was interesting and understandable to people with such diverse backgrounds and professional goals. I have aimed at an audience of professional mathematicians who want to learn some functional analysis, and second-year graduate students who are taking a course in the subject. The only background material needed is what is usually covered in a one-year graduate level course in analysis, and an acquaintance with linear algebra. The book is designed to enable the reader to get actively involved in the development of the mathematics. This can be done by working the starred problems that appear at the end of nearly every section. I often refer to these exercises during subsequent discussions and proofs. Solutions to those starred problems appearing in the introductory chapters (Chapters 1–4) can be found in Appendix A.

The introductory chapters contain the basic facts from the theory of normed spaces. Here the mathematics is developed through the discussion of a sequence of gradually more sophisticated questions. We begin with the most naïve approach of all. In Sections 2 and 3 of Chapter 1, we study finite dimensional normed spaces and ask which of our results are true in the infinite dimensional case. Of course this approach does not lead very far, but it does guide us to some useful facts. In order to carry our discussion of normed spaces further, we take a hint from the history of the subject and learn something about integral equations. This is done in Chapter 2, where we also discuss the Riesz theory of compact operators. A key result in that theory is the theorem associating to each compact operator a pair of complementary subspaces. At this point we inquire into the connection between such pairs of subspaces and continuous projection operators and ask if every closed linear subspace of a normed space has a complement. The discussion of these questions, which occupies some of Chapter 2 and most of Chapter 3, leads us to some very deep theorems. It also exhibits the importance of continuous, linear functionals.

Chapter 4 deals with the weak topology of a normed space, and it also contains an introduction to the theory of locally convex spaces. The latter material is used to prove that a Banach space whose unit ball is compact for the weak topology is reflexive. It is used again in Chapter 5 and in Chapter 7.

One advantage of the approach sketched above is that the important theorems stand out as those which must be appealed to again and again to answer our questions. It should also be mentioned that several of the questions discussed in the text have been the subject of a great many research papers. I have made no attempt to give an account of all of this work. However, the closely related problems of characterizing reflexive Banach spaces and characterizing those Banach spaces that are dual spaces are discussed further in Appendix B.

The last three chapters of the book are independent of one another, and each deals with a special topic. In Chapter 5, John Kelley's elegant proof of the Krien–Milman theorem is presented. That theorem is used to settle the question, Is every Banach space the dual of some other Banach space? (See Chapter 5 for a more precise statement.) Chapter 5 also contains the theorem of Eberlein. I have presented Eberlein's original proof of his famous theorem because I feel that it gives insights into this result not found in more modern proofs. It does not yield the most general result known; but, I feel, the gain in insight is well worth the slight loss in generality. Chapter 6 contains a sample of the interesting, and sometimes surprising, ways that functional analysis enters into discussions of classical analysis. This material can be read immediately after Section 1 of Chapter 4. Distributions are discussed in the last chapter. The Fourier transform is treated early (Section 3) because it requires less machinery than some of the other topics. However, Fourier transforms are not used in any subsequent section. Applications of the theory of distributions to harmonic analysis (Section 3) and to partial differential equations (Section 5e) are also discussed. Readers who are interested only in distributions can read Chapter 7 immediately. They will however occasionally have to go back to Chapter 4 and read some background material.

I would like to take this opportunity to thank Andrea Blum for writing Appendix A. She patiently solved each of these problems and proved that they really can be done. I discussed my ideas for a book with R. P. Boas of Northwestern University and John S. Lomont of the University of Arizona. They each made valuable suggestions, and it is my pleasure now to thank them both. I would also like to thank Louise Fields for the excellent job she did typing the final version of the manuscript.

Remarks on Notation. The chapters are divided into sections. If in a discussion, in say Chapter 4, I want to refer to Theorem 1 in Section 3 of that same chapter, I write "Section 3, Theorem 1." If in that same discussion I want to refer to Theorem 1 in Section 3 of some other chapter, say Chapter 1, then rather than say "Theorem 1 of Section 3 to Chapter I," I simply write "Section 1.3, Theorem 1."

CHAPTER **1**

Preliminaries

Our treatment of functional analysis begins with a long discussion of normed vector spaces. The two most important classes of normed spaces are the Hilbert spaces and the Banach spaces. Although every Hilbert space is a Banach space, the two classes are always treated separately. The concept of a Hilbert space has its origin in the papers of David Hilbert on the theory of integral equations, and it is well known that Hilbert was attracted to this area by the pioneering work of I. Fredholm. Hilbert space theory, by which I mean not only the study of these spaces but also of the continuous, linear operators on them, is one of the most important branches of functional analysis. However, to do the subject justice, we would have to double the length of this book and so, except for an occasional remark, we shall say no more about it.

Stefan Banach was not the first mathematician to investigate the spaces that now bear his name. He did, however, make many important contributions to their study. His book, "Theory of Linear Operations" [2], contains much of what was known about these spaces at the time of its publication (1932), and some of the deepest results in the book are due to Banach himself. The strange title is explained in the preface. Banach writes: "The theory of operations, created by V. Volterra, has as its object the study of functions defined on infinite dimensional spaces." He goes on to discuss the importance of this theory and some of its applications. Now Volterra is recognized, along with Fredholm, as one of the founders of the modern theory of integral equations, and it was undoubtedly his work in this area that led him to the theory of operations.

One fact is clear from our epigrammatic sketch of the history of the theory of normed spaces, and that is that a large portion of this theory has its roots in the study of integral equations. This fact is worth keeping in mind.

1.   Norms on a Vector Space

We shall begin our formal discussion of normed spaces here. In all that follows R and C will denote the field of real numbers and the field of complex numbers, respectively. We shall always assume that these two fields have their standard, metric, topologies, and all of the vector spaces that we consider will be defined over one or the other of these two fields. Sometimes it is not necessary to specify over which of these fields a certain vector space is defined. In that case we shall speak of a vector space over K.

Definition 1.   Let E be a vector space over the field K. A nonnegative, real-valued function p on E is said to be a norm on E if:

(a) $p(x + y) \leq p(x) + p(y)$ for all x, and y in E;
(b) $p(\alpha x) = |\alpha| p(x)$ for all x in E and all α in K;
(c) $p(x) = 0$ if, and only if, $x = 0$.

If p is a norm on E it is customary to denote, for each x in E, the number $p(x)$ by $\|x\|$. A vector space E on which a norm is defined will be called a normed space. If we want to emphasize the norm, say $\|\cdot\|$, on E, we shall speak of the normed space $(E, \|\cdot\|)$.

There is a natural metric associated with $(E, \|\cdot\|)$; we define the distance between any two points x and y of E to be $\|x - y\|$. This metric gives us a topology on E that we call the norm topology of $(E, \|\cdot\|)$, or the topology induced on E by $\|\cdot\|$. Whenever we speak of, say, a norm convergent sequence in E, or a convergent sequence in $(E, \|\cdot\|)$, we mean a sequence of points of E that is convergent for the metric topology induced on E by $\|\cdot\|$. Similarly, we shall speak of norm compact subsets of E (or compact subsets of $(E, \|\cdot\|)$), of norm continuous functions on E, etc.

The plane (i.e., the vector space over R of all ordered pairs of real numbers) with $\|(x, y)\|$ defined to be the square root of $x^2 + y^2$ is, perhaps, the most familiar example of a normed space. More generally, for any fixed, positive integer n, the vector space over K of all ordered n-tuples of elements of K (we shall call it K^n) can be given a norm by defining $\|(x_1, x_2, \ldots, x_n)\|$ to be the square root of $\sum_{j=1}^{n} |x_j|^2$. This will be called the Euclidean norm on K^n.

It is easy to see that, if there is one norm on a vector space E, then there are infinitely many of them; for if $\|\cdot\|$ is a norm on E then so is $\|\cdot\|_\lambda$ where, for the fixed positive real number λ and each x in E, we define $\|x\|_\lambda$ to be $\lambda\|x\|$. There may be other norms on $(E, \|\cdot\|)$ besides those that can be obtained from $\|\cdot\|$ in this way. On the vector space R^2, for example, we have defined $\|(x, y)\|$ to be the square root of $x^2 + y^2$. But we could also define a norm on this space by taking $\|(x, y)\|_1$ to be $|x| + |y|$, or by taking $\|(x, y)\|_2$ to be the maximum of the numbers $|x|$, $|y|$. Now in the applications of the theory of normed spaces one is often concerned with a family of continuous, linear operators on a specific normed space. Here we are using the term "linear operator" to mean a linear map from a vector space into itself. Clearly the set of all continuous, linear operators on a given normed space is determined by the topology on that space and not by the particular norm that induces that topology. So it seems reasonable to say that two norms on a vector space are equivalent if they induce the same topology on the space. This idea is worth further discussion.

Given a normed space $(E, \|\cdot\|)$ and a point x_0 in E, the map $\phi(x) = x + x_0$ is clearly onto, one-to-one, and continuous. It also has a continuous inverse: $\phi^{-1}(x) = x - x_0$. So ϕ is a homeomorphism from E with the norm topology onto itself. Since the point x_0 is arbitrary, this means that the neighborhoods of any point of E are just translates of the neighborhoods of zero. Hence we may compare the topologies induced on E by two different norms by just comparing the systems of neighborhoods of zero in these two topologies.

Definition 2. Let $(E, \|\cdot\|)$ be a normed space. The set $\{x$ in $E \mid \|x\| \le 1\}$ is called the unit ball of E. We shall denote it by $\mathcal{B}_1$ or $\mathcal{B}_1(\|\cdot\|)$.

If $r\mathcal{B}_1$ is taken to mean $\{rx \mid x$ in $\mathcal{B}_1\}$, then any neighborhood of zero contains some set in the family $\{r\mathcal{B}_1 \mid r$ in $R, r > 0\}$.

Definition 3. Let E be a vector space over K and let $\|\cdot\|_1$, $\|\cdot\|_2$ be two norms on E. We shall say that $\|\cdot\|_1$ is weaker than $\|\cdot\|_2$, and we shall write $\|\cdot\|_1 \le \|\cdot\|_2$, if there is a positive number λ such that $\lambda\mathcal{B}_1(\|\cdot\|_2) \subset \mathcal{B}_1(\|\cdot\|_1)$. We shall say that $\|\cdot\|_1$ and $\|\cdot\|_2$ are equivalent, and we shall write $\|\cdot\|_1 \equiv \|\cdot\|_2$, if we have both $\|\cdot\|_2 \le \|\cdot\|_1$ and $\|\cdot\|_1 \le \|\cdot\|_2$.

Suppose that $\|\cdot\|_1$, $\|\cdot\|_2$ are two norms on E with $\|\cdot\|_1 \le \|\cdot\|_2$. Let λ be a positive number such that $\lambda\mathcal{B}_1(\|\cdot\|_2) \subset \mathcal{B}_1(\|\cdot\|_1)$. For any non-zero vector x in E we have $\|\lambda x\|x\|_2^{-1}\|_1 \le 1$. Hence $\lambda\|x\|_1 \le \|x\|_2$ for all nonzero elements of E, and clearly this holds for the zero vector also. Thus we can state: *Two norms $\|\cdot\|_1$ and $\|\cdot\|_2$ on a vector space E are equivalent iff there are positive constants λ and μ such that $\lambda\|x\|_1 \le \|x\|_2 \le \mu\|x\|_1$ for every x in E.*

The identity map on a vector space E, I_E, is defined by the equation $I_E x = x$ for all x in E. This map is, of course, an isomorphism from the vector space E onto itself. If $\|\cdot\|_1$ and $\|\cdot\|_2$ are two norms on E, then the discussion above shows that $\|\cdot\|_1$ is weaker than $\|\cdot\|_2$ iff the map I_E is continuous from $(E, \|\cdot\|_2)$ onto $(E, \|\cdot\|_1)$, and that $\|\cdot\|_1$ is equivalent to $\|\cdot\|_2$ iff I_E is a homeomorphism between these two spaces.

Definition 4. Let $(E_1, \|\cdot\|_1)$ and $(E_2, \|\cdot\|_2)$ be two normed spaces over the same field. Let T be an isomorphism from E_1 onto E_2. We shall say that T is a topological isomorphism if it is a homeomorphism from $(E_1, \|\cdot\|_1)$ onto $(E_2, \|\cdot\|_2)$.

We note that two norms $\|\cdot\|_1$, $\|\cdot\|_2$ on a vector space E are equivalent iff the identity map is a topological isomorphism from $(E, \|\cdot\|_1)$ onto $(E, \|\cdot\|_2)$.

EXERCISES 1

A number of useful facts, facts that will be referred to later on in the text, are scattered among the exercises. Any problem that is referred to later on is marked with a star. The number of starred problems will decrease as the material gets more difficult.

*1. Let $(E, \|\cdot\|)$ be a normed space, let $\{x_n\}$ be a sequence of points of E, and suppose that this sequence converges to a point x_0 of E for the norm topology. Show that $\lim \|x_n\| = \|x_0\|$.

*2. Let E, F be two normed spaces over the same field and let u be a linear map from E into F.
  (a) Show that u is continuous on E iff it is continuous at zero.
  (b) Show that u is continuous on E iff there is a constant M such that $\|u(x)\| \leq M$ for all x in the unit ball of E. Hint: If u is continuous at zero but no such M exists, then for each positive integer n we can find a point x_n in E such that $\|x_n\| \leq 1$ and $\|u(x_n)\| \geq n$.
  (c) Show that u is continuous on E iff there is a constant M such that $\|u(x)\| \leq M\|x\|$ for all x in E.

3. We have defined three different norms on the vector space R^2. Sketch the unit ball of each of these three normed spaces. Show that any two of our three norms are equivalent.

*4. Let E, F be two normed spaces over the same field and let u be a topological isomorphism from E onto F. Denote the norm on E by $\|\cdot\|_E$ and the norm on F by $\|\cdot\|_F$.
  (a) For each x in E define $|\|x\||$ to be $\|u(x)\|_F$. Show that $|\|\cdot\||$ is a norm on E and that it is equivalent to $\|\cdot\|_E$.
  (b) For each y in F define $|\|y\||$ as follows: Let x be the unique element of E such that $u(x) = y$ and take $|\|y\||$ to be $\|x\|_E$. Show that $|\|\cdot\||$ is a norm on F and that it is equivalent to $\|\cdot\|_F$.
  (c) Suppose now that E is just a vector space over K, F is a normed space over K, and u is an isomorphism from E onto F. The function $|\|\cdot\||$ defined on E as in (a) is still a norm on E. Show that if E is given this norm then u becomes a topological isomorphism from E onto F. If E is a normed space and F is just a vector space then similar remarks can be made about the function defined on F as in (b).

∗5.   Let E, F be two normed spaces over the same field and let u be a topological isomorphism from E onto F. If $|\cdot|_E$ and $|\cdot|_F$ are norms on E and F, respectively, that are equivalent to the original norms on these spaces, show that u is a topological isomorphism from $(E, |\cdot|_E)$ onto $(F, |\cdot|_F)$.

2.   Finite Dimensional Normed Spaces

Before continuing with our general discussion it is instructive to investigate the properties of a special class of normed spaces. We have in mind spaces $(E, \|\cdot\|)$ for which E is a finite dimensional vector space over K. Such spaces do arise in applications.

Recall that a vector space over K is said to be finite dimensional if, for some nonnegative integer n, it has a basis containing n elements; the number n is called the dimension of the space. We allow n to be nonnegative in order to include the trivial vector space, i.e., the vector space over K whose only element is the zero vector. A basis for this space is, by convention, the empty set. Hence the trivial vector space has dimension zero. It is easy to see that any finite dimensional vector space over K can be given a norm. In fact

Theorem 1.   Any two norms on a finite dimensional vector space are equivalent.

Proof.   Let F be a finite dimensional vector space over K and let $\|\cdot\|_1$ and $\|\cdot\|_2$ be two norms on F. Choose a basis $x_1, x_2, \ldots, x_n$ for F and define a third norm, $|\cdot|$, as follows: For each x in F there is a unique set of scalars $\alpha_1, \alpha_2, \ldots, \alpha_n$ such that $x = \sum \alpha_j x_j$. Take $|x|$ to be the maximum of the numbers $|\alpha_1|, |\alpha_2|, \ldots, |\alpha_n|$. Suppose that each of the norms $\|\cdot\|_1$ and $\|\cdot\|_2$ is equivalent to $|\cdot|$. Then there are positive scalars m_1, M_1 and m_2, M_2 such that

$$m_1 |x| \le \|x\|_1 \le M_1 |x| \qquad \text{and} \qquad m_2|x| \le \|x\|_2 \le M_2|x|$$

for each x in F. It follows that

$$(m_2/M_1)\|x\|_1 \le m_2|x| \le \|x\|_2 \le M_2|x| \le (M_2/m_1)\|x\|_1$$

for each x in F, and hence $\|\cdot\|_1$ and $\|\cdot\|_2$ are equivalent.

Now let $\|\cdot\|$ denote either $\|\cdot\|_1$ or $\|\cdot\|_2$. We shall show that $\|\cdot\|$ is equivalent to $|\cdot|$. If $x = \sum \alpha_j x_j$ then

$$\|x\| \leq \sum |\alpha_j| \|x_j\| \leq |x| (\sum \|x_j\|).$$

Since $(\sum \|x_j\|)$ is a constant we see that $\|\cdot\|$ is weaker than $|\cdot|$ on F. Let $S = \{x \text{ in } F \mid |x| = 1\}$ and choose a sequence $\{y_k\}$ of points of S such that $\lim \|y_k\| = \inf\{\|x\| \mid x \text{ in } S\}$. For each y_k there are scalars α_{k1}, $\alpha_{k2}, \ldots, \alpha_{kn}$ such that $y_k = \sum \alpha_{kj} x_j$. Since each $y_k \in S$, $|\alpha_{kj}| \leq 1$ for $j = 1, 2, \ldots, n$ and every k. These inequalities imply that there is a subsequence of $\{y_k\}$ (call it $\{y_k\}$ also) such that $\lim \alpha_{kj}$ exists and equals, say, α_j for $j = 1, 2, \ldots, n$. Let $y = \sum \alpha_j x_j$ and note that $\{y_k\}$ converges to y for $|\cdot|$, i.e., $\lim |y - y_k| = 0$. It follows that $|y| = 1$ (Exercises 1, problem 1) and hence, in particular, $y \neq 0$.

At this point we make an observation. Let λ be the maximum of the numbers $\|x_j\|$, $j = 1, 2, \ldots, n$. If a positive number ε is given then the elements $x = \sum \beta_j x_j$ and $z = \sum \gamma_j x_j$ of F will satisfy the inequality $\|x - z\| < \varepsilon$, if $|\beta_j - \gamma_j| < \varepsilon/\lambda n$ for each j, i.e., if $|x - z| < \varepsilon/\lambda n$. This fact, together with the remarks contained in the last paragraph, implies that $\lim \|y_k\| = \|y\|$. So $\|y\| = \inf\{\|x\| \mid x \text{ in } S\}$ and since $y \neq 0$ this infimum is positive. Now if x is any nonzero element of F, then $x/|x|$ is in S and hence $\|x\| \geq \|y\| |x|$. It follows that $|\cdot|$ is weaker than $\|\cdot\|$ on F.

Corollary 1. If two finite dimensional normed spaces over K have the same dimension, then they are topologically isomorphic.

Proof. It suffices to show that if a normed space $(F, \|\cdot\|)$ over K has (finite) dimension n then it is topologically isomorphic to the space K^n with the Euclidean norm. There is an isomorphism u from F onto K^n. We can use this map to define a new norm on F as follows: For each x in F let $|\|x\||$ be the norm of $u(x)$ in K^n (Exercises 1, problem 4a). When F is given this new norm u becomes a topological isomorphism (Exercises 1, problem 4c). However, Theorem 1 shows that $|\|\cdot\||$ is equivalent to $\|\cdot\|$ on F. Hence u is a topological isomorphism from $(F, \|\cdot\|)$ onto K^n (Exercises 1, problem 5).

Theorem 1 has two other useful corollaries. In order to state them we need some more terminology. The proofs of these corollaries will be left to the reader (see problem 1 below). We have already noted that a normed space $(E, \|\cdot\|)$ has an associated metric. If E, with this metric, is a complete metric space then we shall say that $(E, \|\cdot\|)$ is a complete normed space. More explicitly:

Definition 1.   Let $(E, \|\cdot\|)$ be a normed space and let $\{x_n\}$ be a sequence of points of E. We shall say $\{x_n\}$ is a Cauchy sequence if the limit, as m and n tend to infinity, of $\|x_n - x_m\|$ is zero. We shall say that $(E, \|\cdot\|)$ is a complete normed space, or that $(E, \|\cdot\|)$ is a Banach space, if every Cauchy sequence of points of E converges to a point of E.

Corollary 2.   Any finite dimensional normed space is a Banach space.

If G is a linear subspace of a normed space $(E, \|\cdot\|)$, then G can be given a norm in a natural way; simply regard each element of G as an element of E and take its norm in E. More formally we have an inclusion map j from G into E, $j(x) = x$ for each x in G, and we define the norm of $x \in G$ to be $\|j(x)\|$. This is called the subspace norm on G. Whenever we work with a linear subspace of a normed space we shall always assume, without explicit mention, that it has the subspace norm.

Corollary 3.   Any finite dimensional linear subspace of a normed space E is a closed subset of E.

A linear subspace of a normed space E that is also a closed subset of E will be called a closed, linear subspace of E.

We know that any closed, bounded subset of, say, R^2 is compact. This is also true for each of the spaces K^n. Is anything like this true for normed spaces? This question will lead us to one of the fundamental differences between finite dimensional and infinite dimensional normed spaces.

Definition 2.   Let $(E, \|\cdot\|)$ be a normed space and let S be a subset of E. We shall say that S is a bounded subset of E if there is a positive scalar λ such that $S \subset \lambda \mathscr{B}_1$.

It is clear that a normed space has the property that each of its closed, bounded subsets is compact iff the unit ball of the space is compact.

Theorem 2.   The unit ball of a normed space is compact iff the space is finite dimensional.

Proof. The sufficiency of our condition follows from Corollary 1. Let $(F, \|\cdot\|)$ be a normed space over K and assume that the unit ball of this space is compact. If F is not finite dimensional then we can find a sequence $\{x_n\}$ of points of F such that any finite subset of $\{x_n\}$ is linearly independent. For each positive integer k let F_k be the linear subspace of F that is generated by $x_1, x_2, \ldots, x_k$. Note that, by Corollary 3, F_k is a closed, linear subspace of F_{k+1} for each k. Suppose that we can choose, for each k, a point y_k in F_{k+1} such that $\|y_k\| = 1$ and $\|x - y_k\| \geq \frac{1}{2}$ for all x in F_k. Then $\{y_k\}$ is contained in the unit ball of $(F, \|\cdot\|)$ and, since $\|y_k - y_{k+1}\| \geq \frac{1}{2}$ for every k, no subsequence of $\{y_k\}$ is convergent. This contradiction shows that F must be finite dimensional. We shall now show that we can choose an element y_k with the properties stated above.

Lemma. Let E be a normed space over K, let G be a closed, linear subspace of E and let δ be a real number; $0 \leq \delta < 1$. If $G \neq E$, then there is a point y_0 in E such that $\|y_0\| = 1$ and $\|x - y_0\| \geq \delta$ for all x in G.

Proof. Choose y in E, y not in G, and let d be the positive number $\inf\{\|y - x\| \,|\, x \text{ in } G\}$. For any fixed point x_0 in G we may write

$$\|x - (y - x_0)\|y - x_0\|^{-1}\| = \|\{x\|y - x_0\| - (y - x_0)\}\|y - x_0\|^{-1}\|$$

$$= \|\{(x\|y - x_0\| + x_0) - y\}\|y - x_0\|^{-1}\|.$$

This last term is greater than or equal to $d\|y - x_0\|^{-1}$ for every x in G. The point y_0 may be found by first taking x_0 in G such that $\|y - x_0\| < d\delta^{-1}$, and then setting y_0 equal to $(y - x_0)\|y - x_0\|^{-1}$.

We shall give many examples of normed spaces in the following pages. In all of these examples we begin with a family of functions each member of which is defined on the same set, say S, and takes its values in K or in some fixed vector space over K. Such a family will be called a space of functions on S if it is a vector space with respect to the following operations: For any f, g in the family and any α in K, $(f + g)(s) = f(s) + g(s)$ for each s in S and $\alpha f(s) = \alpha[f(s)]$ for each s in S. Thus if S is, for example, $[0, 1] = \{x \text{ in } R \mid 0 \leq x \leq 1\}$, then the family of all K-valued, continuous functions on S is a space of functions on S.

EXERCISES 2

∗1. Prove Corollaries 2 and 3 to Theorem 1. Prove the statement made just after Definition 2. Prove that a Cauchy sequence in a normed space $(E, \|\cdot\|)$ is a bounded set.

∗2. Let E be a normed space over K. For any subset S of E let cl S denote the closure of S.
  (a) If H is a linear subspace of E prove that cl H is a closed, linear subspace of E.
  (b) If B is any bounded subset of E prove that cl B is a bounded subset of E.

∗3. Let E, F be two normed spaces over the same field and let u be a linear map from E into F.
  (a) Prove that u is continuous on E iff the set $u(B) = \{u(x) \mid x \text{ in } B\}$ is bounded in F whenever the set B is bounded in E.
  (b) If E is a finite dimensional normed space show that u is continuous.

4. Let l_∞ denote the space of all sequences $\{x_n\}$ of points of K such that $\sup\{|x_n| \mid n = 1, 2, \ldots\}$ is finite.
  ∗(a) Show that l_∞ is an infinite dimensional space of functions on the positive integers.
  ∗(b) For each $x = \{x_n\} \in l_\infty$ define $\|x\|_\infty$ to be $\sup\{|x_n| \mid n = 1, 2, \ldots\}$. Show that this function is a norm on l_∞.
  (c) Let $F = \{x = \{x_n\} \text{ in } l_\infty \mid x_n = 0 \text{ except for finitely many } n\}$. Show that F is a linear subspace of l_∞. What is the closure of F in l_∞?
  (d) Referring to Corollaries 2 and 3 what can you say about:
    (i) the dimension of F;
    (ii) the vector space F with the subspace norm?

5. Let l_1 denote the space of all sequences $\{x_n\}$ of points of K such that $\sum_{n=1}^{\infty} |x_n| < \infty$.
  ∗(a) Show that l_1 is an infinite dimensional space of functions on the positive integers.

*(b) For each $x = \{x_n\}$ in l_1 define $\|x\|_1$ to be $\sum_{n=1}^{\infty} |x_n|$. Show that this function is a norm on l_1.

(c) Let $F = \{x = \{x_n\}$ in $l_1 \,|\, x_n = 0$ except for finitely many $n\}$. Show that F is a linear subspace of l_1. What is the closure of F in l_1?

(d) Referring to Corollary 2 what can you say about F with the subspace norm?

3. Infinite Dimensional Spaces.
Hamel and Schauder Bases

In our treatment of finite dimensional normed spaces we made repeated use of the fact that any such space has a basis. Let us take a look now at some of the generalizations of this concept.

Definition 1. Let X be a vector space over K. A subset $\mathscr{H}$ of X is said to be a Hamel basis for X if it has the two following properties:

(a) Any finite subset of $\mathscr{H}$ is linearly independent.

(b) Given any x in X there is a finite subset $h_1, h_2, \ldots, h_n$ of $\mathscr{H}$ such that $x, h_1, h_2, \ldots, h_n$ is linearly dependent.

It is an easy application of Zorn's lemma [15, Theorem 25(e), p. 33] to prove that any linearly independent subset of X is contained in a Hamel basis for X. In particular, any vector space over K has a Hamel basis. Observe that, by (b), every element of X is a linear combination of some finite subset of $\mathscr{H}$. Furthermore, if we require that the coefficients in such a linear combination be nonzero, then, by (a), both the finite set and the coefficients are uniquely determined by x. So for each x in X we can find a unique finite set $h_1, \ldots, h_n$ in $\mathscr{H}$ and a unique set of nonzero elements $\alpha_1, \alpha_2, \ldots, \alpha_n$ of K such that $x = \sum_{j=1}^{n} \alpha_j h_j$. Define $\|x\,|\,\mathscr{H}\|$ to be the maximum of the numbers $|\alpha_j|, j = 1, 2, \ldots, n$. Clearly $\|x\,|\,\mathscr{H}\|$ is the norm for X and we have proved: *Any vector space over K can be given a norm.*

There is one important space for which a Hamel basis can be exhibited. A K-valued function f on $[0, 1]$ is said to be a polynomial function if there is a nonnegative integer n and elements $a_0, a_1, \ldots, a_n$ of K such that $f(x) = \sum_{j=0}^{n} a_j x^j$ for all x in $[0, 1]$. The set of all polynomial functions on $[0, 1]$ is a space of functions on $[0, 1]$ that we shall denote by $\mathscr{P}[0, 1]$. Clearly $\{1, x, x^2, \ldots\}$ is a Hamel basis for $\mathscr{P}[0, 1]$.

The set of all K-valued, continuous functions on $[0, 1]$ is a space of functions, which we will denote by $\mathscr{C}[0, 1]$. We can define a norm for this space as follows: For each f in $\mathscr{C}[0, 1]$ let $\|f\|_\infty$ be the maximum of the numbers $\{|f(x)| \,|\, 0 \le x \le 1\}$. Observe that a sequence of continuous functions is convergent for this norm iff it is uniformly convergent on $[0, 1]$. Hence $(\mathscr{C}[0, 1], \|\cdot\|_\infty)$ is a Banach space. Notice that, in contrast to the finite dimensional case, $\mathscr{P}[0, 1]$ is a linear subspace of $\mathscr{C}[0, 1]$ that is dense in this space for the norm topology; this follows from the Weierstrass approximation theorem [10, Corollary 7.31, p. 96]. We can use the existence of such a subspace to prove that two norms on an infinite dimensional space need not be equivalent.

Let $\mathscr{H}_1$ be a Hamel basis for $\mathscr{P}[0, 1]$. We can find a Hamel basis $\mathscr{H}$ for $\mathscr{C}[0, 1]$ that contains $\mathscr{H}_1$. Choose any point h_0 that is in $\mathscr{H}$ but not in $\mathscr{H}_1$ and define $\phi_0(h_0) = 1$, $\phi_0(h) = 0$ for all h in $\mathscr{H}$, $h \ne h_0$. Then extend ϕ_0 to all of $\mathscr{C}[0, 1]$ by requiring that it be linear. It is clear that ϕ_0 is a continuous, linear map from $(\mathscr{C}[0, 1], \|\cdot|\mathscr{H}\|)$ into K because

$$|\phi_0(f)| = \left|\phi_0\left(\sum_{j=1}^n \alpha_j h_j\right)\right| \le \sum |\alpha_j| |\phi_0(h_j)|$$

$$\le \max\{|\alpha_j| \,|\, 1 \le j \le n\} \sum |\phi_0(h_j)| \le \|f\|\mathscr{H}\|$$

for each f in $\mathscr{C}[0, 1]$. If ϕ_0 were continuous on $(\mathscr{C}[0, 1], \|\cdot\|_\infty)$, then $\{f \,|\, \phi_0(f) = 0\}$ would be a closed, linear subspace of $(\mathscr{C}[0, 1], \|\cdot\|_\infty)$. But this last set contains $\mathscr{P}[0, 1]$, which is dense in $(\mathscr{C}[0, 1], \|\cdot\|_\infty)$. Hence if ϕ_0 were continuous on $(C[0, 1], \|\cdot\|_\infty)$, it would have to be identically zero, contradicting the fact that $\phi_0(h_0) \ne 0$. Observe that $\|\cdot|\mathscr{H}\|$ and $\|\cdot\|_\infty$ could not be equivalent norms on $\mathscr{C}[0, 1]$.

We have just seen that a nonzero linear map, from a normed space E into the underlying field, whose null space is dense in E must be discontinuous. The converse is also true. Before proving it let us give these maps a name.

Definition 2.   Let X be a vector space over K. A linear map from X into K is called a linear functional on X. The set of all linear functionals on X will be denoted by $X^\#$.

It is clear that $X^\#$ is a space of functions on X. We shall call it the algebraic dual space of X or, more simply, the algebraic dual of X. If $(E, \|\cdot\|)$ is a normed space over K, then an element ϕ of E^{*} is continuous on $(E, \|\cdot\|)$ iff it is bounded on the unit ball of E (Exercises 1, problem 2b). If ϕ is not continuous then $\{\phi(x) \mid x$ in E, $\|x\| \leq 1\}$ is all of K. This is obvious if $K = R$. The proof, when $K = C$, is left to the reader (see problem 6 below).

Lemma 1.   Let $(E, \|\cdot\|)$ be a normed space over K and let ϕ be a nonzero element of E^{*}. Then the following conditions on ϕ are equivalent:

(a) ϕ is continuous on $(E, \|\cdot\|)$.
(b)   The null space of ϕ is a proper, closed linear subspace of $(E, \|\cdot\|)$.
(c)   The null space of ϕ is not dense in $(E, \|\cdot\|)$.

Proof. We need only show that (c) implies (a). Let $N(\phi)$ be the null space of ϕ. If we assume that (c) is true, then there is an open set G in E that is nonempty and does not meet $N(\phi)$. From the discussion in Section 1 there is, for each point x of G, a positive scalar λ such that $\{x + \lambda y \mid y$ in $\mathscr{B}_1\}$ is disjoint from $N(\phi)$. If ϕ is not continuous then $\{\phi(\lambda y) \mid y$ in $\mathscr{B}_1\}$ is all of K. In particular, for some y_0 in $\mathscr{B}_1$, $\phi(\lambda y_0) = -\phi(x)$. But then $x + \lambda y_0$ is in $N(\phi)$ and we have reached a contradiction.

There is another type of basis that has proved itself useful in the study of certain normed spaces. For various reasons, which need not concern us at the moment, one usually defines this type of basis only for a Banach space.

Definition 3.   Let $(B, \|\cdot\|)$ be a Banach space over K. A sequence $\{x_n\}$ of points of B is said to be a Schauder basis for B if for each element x in B there is a unique sequence $\{a_n\}$ of points of K such that

$$\lim_{n \to \infty} \left\| x - \sum_{j=1}^{n} a_j x_j \right\| = 0.$$

A Banach space need not have a Schauder basis. If $(B, \|\cdot\|)$ is a Banach space over R and $\{x_n\}$ is a Schauder basis for B, then the countable set $\{\sum_{j=1}^{n} q_j x_j \mid n$ is a positive integer and q_j, $1 \leq j \leq n$, are arbitrary rational numbers} is dense in $(B, \|\cdot\|)$. Similarly, a Banach space over C that has a Schauder basis must contain a countable dense set. Hence, if a Banach space is to have a Schauder basis, then it must have the following property.

Definition 4. A normed space $(E, \|\cdot\|)$ is said to be a separable normed space if E has a countable subset that is dense in E for the norm topology.

It is useful to notice that any subset of a separable normed space is separable, i.e., if $(E, \|\cdot\|)$ is a separable normed space and if S is any subset of E, then S contains a countable set whose closure in E includes all of S [21, Proposition 13, p. 138].

We have seen that only a separable Banach space can have a Schauder basis. For many years it was not known whether every separable Banach space did in fact have such a basis. The question was finally settled, in the negative, by Enflo [6]. There is an extensive literature on the subject of Schauder bases (see [19] and [22]). However, this subject is, we feel, too specialized to be included in this text and, except for the results stated in problems 2 and 3 (which we will use in Chapter 3), we will make no further reference to it.

The space $(l_1, \|\cdot\|)$, defined in problem 5 of Exercises 2, will be referred to later on. Let us show that this is a Banach space. Let $\{v^n \,|\, n = 1, 2, \ldots\}$ be a Cauchy sequence in this space; here $v^n = \{x_j^n \,|\, j = 1, 2, \ldots\}$ for each n. Then if $\varepsilon > 0$ is given there is an integer N such that $\|v^n - v^m\|_1 < \varepsilon$ whenever both m and n are $\geq N$, i.e., $\sum_{j=1}^{\infty} |x_j^n - x_j^m| < \varepsilon$ whenever both m and n are $\geq N$. Two things follow from this:

(a) For each fixed j the sequence $\{x_j^n \,|\, n = 1, 2, \ldots\}$ is a Cauchy sequence in the field K.

(b) For each fixed integer k, $\sum_{j=1}^{k} |x_j^n - x_j^m| < \varepsilon$ whenever both m and n are $\geq N$.

For each j let $x_j \in K$ be the limit of $\{x_j^n \,|\, n = 1, 2, \ldots\}$. We shall show that $v = \{x_j \,|\, j = 1, 2, \ldots\}$ is in l_1. To do this, for any fixed k let $v_k = (x_1, x_2, \ldots, x_k, 0, \ldots)$ and let $v_k^n = (x_1^n, x_2^n, \ldots, x_k^n, 0, 0, \ldots)$ for each n. Clearly v_k and each v_k^n is in l_1. Hence

$$\|v_k\|_1 \leq \|v_k - v_k^n\|_1 + \|v_k^n\|_1 \leq \varepsilon + \sum_{j=1}^{k} |x_j^n|$$

whenever $n \geq N$. If we fix $n \geq N$ and let k tend to infinity, we find that $\sum_{j=1}^{\infty} |x_j| \leq \varepsilon + \|v^n\|_1$. Hence $v = \{x_j\}$ is in l_1.

The only thing left to prove is that $\lim \|v^n - v\|_1 = 0$. By (b) we see that, for any fixed k, $\sum_{j=1}^{k} |x_j^n - x_j| \le \varepsilon$ whenever $n \ge N$. Holding $n \ge N$ fixed and letting k tend to infinity gives $\|v^n - v\|_1 \le \varepsilon$ whenever $n \ge N$.

EXERCISES 3

1. Show that $(\mathscr{C}[0, 1], \|\cdot\|_\infty)$ is a separable Banach space.

*2. Consider the normed space $(l_\infty, \|\cdot\|_\infty)$ defined in problem 4 of Exercises 2.
   (a) Show that this space is complete. Hint: Refer to the proof that $(l_1, \|\cdot\|_1)$ is complete.
   (b) For each fixed, positive integer n let e_n be the sequence with 1 in the nth place and zero everywhere else. Prove that the set $e_1, e_2, \ldots$ is *not* a Hamel basis for l_∞ and that it is *not* a Schauder basis for l_∞.
   (c) Let $c_0 = \{\{x_n\}$ in $l_\infty | \lim x_n = 0\}$. Show that c_0 is a closed, linear subspace of $(l_\infty, \|\cdot\|_\infty)$ and that $e_1, e_2, \ldots$ is a Schauder basis for c_0.
   (d) Prove that $(l_\infty, \|\cdot\|_\infty)$ can not have a Schauder basis.

*3. Show that the set $e_1, e_2, \ldots$ (see problem *2b) is *not* a Hamel basis for l_1 but that it is a Schauder basis for $(l_1, \|\cdot\|_1)$.

*4. Let X be a vector space over K. A linear subspace Y of X is said to have codimension one in X if the following is true: There is an x_0 in X such that for any x in X we can find y in Y and λ in K, both depending on x, for which $x = y + \lambda x_0$.
   (a) Show that a proper, linear subspace of X has codimension one iff it is the null space of some nonzero element of $X^\#$.
   (b) Let θ, ϕ be nonzero elements of $X^\#$. Show that the null space of θ is contained in the null space of ϕ iff there is a scalar λ such that $\phi = \lambda\theta$.

*5. Let n be a fixed, positive integer and consider the space K^n. If f is any linear functional on K^n show that there is a unique n-tuple $(c_1, c_2, \ldots, c_n)$, each $c_j \in K$, such that
$$f[x] = f[(x_1, \ldots, x_n)] = \sum_{j=1}^{n} c_j x_j$$

for all $x = (x_1, \ldots, x_n)$ in K^n. Hence a linear subspace of codimension one in K^n coincides with the set of all solutions to an equation of the form $c_1 x_1 + c_2 x_2 + \cdots + c_n x_n = 0$, where $c_1, c_2, \ldots, c_n$ are fixed elements of K not all zero.

6. Let $(E, \|\cdot\|)$ be a normed space over C. If $\phi \in E^{\#}$ is not continuous on E, prove that $\{\phi(x) \mid x$ in E, $\|x\| \leq 1\}$ is all of C. Hint: If $x_0 \in E$, $\|x_0\| \leq 1$ then $e^{i\theta}x_0$ is also in E, $\|e^{i\theta}x_0\| = \|x_0\|$, and $\phi(e^{i\theta}x_0) = e^{i\theta}\phi(x_0)$.

CHAPTER 2

Operator Theory

From one point of view the study of the finite dimensional normed spaces is rather disappointing. Apart from the few, mostly negative, results mentioned in Section 1.3, it has taught us little about general normed spaces. In fact, since, as we have seen, most of the theorems that are true for this class of spaces are true only for this class it is clear that we have been exploring a kind of blind alley. So, if our investigation of general normed spaces is to proceed any further, we must find another way of approaching them. We have already mentioned the importance of the work on integral equations for the early development of the theory of normed spaces. Let us take a look at an integral equation.

1. Compact Linear Operators

Consider the Banach space $(\mathscr{C}[0, 1], \|\cdot\|_\infty)$ and a function $K(x, t)$ that is continuous for $0 \le t,\ x \le 1$. If an element $g(x) \in \mathscr{C}[0, 1]$ is given, we may ask whether there is an element $u(x)$ of this space such that

$$(*) \qquad\qquad u(x) = g(x) + \int_0^1 K(x, t)u(t) \, dt.$$

Equation $(*)$ is a Fredholm equation of the second kind. Such equations arise in many applications of mathematics [24].

If $f(x) \in \mathscr{C}[0, 1]$ then the function $\int_0^1 K(x, t)f(t) \, dt$ (call it $Kf(x)$) is also in $\mathscr{C}[0, 1]$. Furthermore, if

$$M = \max\{|K(x, t)| \,|\, 0 \le t, x \le 1\}$$

then $|Kf(x)| \le M\|f\|_\infty$ for all x, i.e., $\|Kf\|_\infty \le M\|f\|_\infty$. Thus the map K that takes each $f(x)$ in $\mathscr{C}[0, 1]$ to $Kf(x)$ is a continuous, linear operator on the Banach space $(\mathscr{C}[0, 1], \|\cdot\|_\infty)$. Let I denote the identity operator on our Banach space, and note that the equation $(*)$ has a solution in $\mathscr{C}[0, 1]$ for a given function $g(x)$ iff $g(x)$ lies in the range of $I - K$. Also, solutions to $(*)$, when they exist, will be unique iff the null space of $I - K$ contains only the zero vector. So, to answer questions about the existence and uniqueness of solutions to equation $(*)$ we should investigate the range and the null space of the operator $I - K$. We shall do that, but first let us examine the operator K more closely.

Let $\{h_n(x)\}$ be a bounded sequence in $(\mathscr{C}[0, 1], \|\cdot\|_\infty)$; say $\|h_n(x)\|_\infty \le l$ for all n. Then certainly $\{Kh_n(x)\}$ is a bounded sequence. But more is true. We have, for any fixed n,

$$|Kh_n(x) - Kh_n(y)| \le \int_0^1 |K(x, t) - K(y, t)| \, |h_n(t)| \, dt$$

$$\le l \max\{|K(x, t) - K(y, t)| \, |t\}.$$

Now $K(x, t)$ is uniformly continuous. Thus, given $\varepsilon > 0$ we can find a $\delta > 0$ such that $|Kh_n(x) - Kh_n(y)| < \varepsilon$ whenever $|x - y| < \delta$, and this is true for every n. It follows that $\{Kh_n(x)\}$ is equicontinuous on $[0, 1]$ and so, by the Ascoli–Arzela theorem [21, Theorem 33, p. 179], some subsequence of it converges uniformly (i.e., for $\|\cdot\|_\infty$). So the operator K maps any sequence in $\mathscr{C}[0, 1]$ that is bounded for $\|\cdot\|_\infty$ onto one that has a convergent subsequence for $\|\cdot\|_\infty$.

Definition 1.   Let $(B, \|\cdot\|)$ be a Banach space and let K be a continuous, linear operator on B. We shall say that K is a compact operator on B, or simply that K is compact, if for any sequence $\{x_n\}$ of points of

B that is bounded for $\|\cdot\|$ the sequence $\{Kx_n\}$ has a subsequence that is convergent for $\|\cdot\|$.

It is easy to see that a continuous, linear operator on a Banach space is a compact operator iff it maps any bounded set onto a set whose closure is compact.

We have seen that questions about the existence and uniqueness of solutions to certain integral equations lead to questions about the range and null space of an operator of the form $I - K$, where I is the identity operator and K is a compact operator on a certain Banach space $(B, \|\cdot\|)$. Let these symbols retain this meaning for the rest of this section, and let $N(I - K)$ be the null space of $I - K$.

Theorem 1. The null space of $I - K$ is a finite dimensional subspace of B.

Proof. The null space of $I - K$ is a closed, linear subspace of B. Let $\mathscr{B} = \{x \text{ in } N(I - K) \mid \|x\| \leq 1\}$. Since $N(I - K)$ is $\{y \text{ in } B \mid y = Ky\}$, we must have $K(\mathscr{B}) = \mathscr{B}$. Thus $K(\mathscr{B})$ is a compact set. But $\mathscr{B}$ is the unit ball of the Banach space $N(I - K)$ and so this space must be finite dimensional (Section 1.2, Theorem 2).

Applying Theorem 1 to equation (∗), we see that there is a finite linearly independent set $u_1(x)$, $u_2(x)$, ..., $u_k(x)$ in $\mathscr{C}[0, 1]$ that spans $N(I - K)$, and if $u_g(x)$ is a solution to (∗) for a given function $g(x)$, then the general solution is

$$u(x) = u_g(x) + \sum_{j=1}^{k} \alpha_j u_j(x);$$

$\alpha_1, \alpha_2, \ldots, \alpha_k$ are arbitrary scalars.

If y is in the range of $I - K$, denote this linear subspace of B by $R(I - K)$, then there is an x in B such that $y = (I - K)x$. Even more is true.

Lemma 1. There is a number M with the following property: For each y in $R(I - K)$ there is an x in B such that $y = (I - K)x$ and $\|x\| \leq M\|y\|$.

Proof. Suppose that the lemma is false. Then for each positive integer n there is a point y_n in $R(I - K)$ such that whenever $x_n \in B$

satisfies the equation $y_n = (I - K)x_n$ it must also satisfy the inequality $\|x_n\| > n\|y_n\|$. Clearly $y_n \neq 0$ because the inequality is strict. For $n = 1, 2, \ldots$ choose w_n in B such that $(I - K)w_n = y_n$. Now no w_n is in $N(I - K)$ and this subspace is closed, so each of the numbers $d_n = \inf\{\|w_n - x\| \,|\, x \text{ in } N(I - K)\}$ is positive. Hence for each n we can choose v_n in $N(I - K)$ such that $d_n \le \|w_n - v_n\| < 2d_n$.

Define z_n to be $(w_n - v_n)\|w_n - v_n\|^{-1}$ for $n = 1, 2, \ldots$ Since $\{z_n\}$ is a bounded sequence and K is a compact operator we may assume that $\{Kz_n\}$ is convergent. Now $I - K$ applied to $\|w_n - v_n\|z_n$ is just $(I - K)w_n$, which is y_n. Thus we have $\|w_n - v_n\|\|z_n\| > n\|y_n\|$. It follows that

$$\|(I - K)z_n\| = \|y_n\| \|w_n - v_n\|^{-1} < \|z_n\|/n = 1/n,$$

and so $\lim(I - K)z_n = 0$. Since we can write $z_n = (I - K)z_n + Kz_n$, it is clear that $\lim z_n$ exists. Call this limit z. Then since $(I - K)z = \lim(I - K)z_n = 0$, z is in $N(I - K)$. But

$$\|z_n - z\| = \| \|w_n - v_n\|^{-1}(w_n - v_n) - z\|$$

$$= \|w_n - (v_n + \|w_n - v_n\|z)\| \|w_n - v_n\|^{-1} \ge d_n\|w_n - v_n\|^{-1} \ge \tfrac{1}{2}.$$

This contradicts the fact that $\lim z_n = z$.

Theorem 2.   The range of $I - K$ is a closed linear subspace of B.

Proof.   Let $\{y_n\}$ be a sequence of points of $R(I - K)$ that converges to $y \in B$. Let M be the number whose existence was proved in Lemma 1. For each integer n we can choose x_n in B such that $y_n = (I - K)x_n$ and $\|x_n\| \le M\|y_n\|$. Since $\{y_n\}$ is a convergent sequence it is bounded (Exercises 1.2, problem 1) and so $\{x_n\}$ is bounded. Hence we may assume that $\{Kx_n\}$ is convergent. Clearly $x_n = y_n + Kx_n$, for each n, and so $\{x_n\}$ must converge to some point, say x, of B. But then

$$(I - K)x = \lim(I - K)x_n = \lim y_n = y$$

and so $y \in R(I - K)$.

We will have more to say about the range and null space of $I - K$ in the next section. Let us now return to our integral equation and see how we might go about solving it. The equation reads

$$(*)\qquad\qquad u(x) = g(x) + \int_0^1 K(x, t)u(t)\, dt,$$

or

$$u(x) = g(x) + Ku(x).$$

Suppose that $g(x) \in \mathscr{C}[0, 1]$ is given and construct a sequence in $\mathscr{C}[0, 1]$ as follows: Take $f_0(x) \equiv 0$, $f_1(x) = g(x) + Kf_0(x)$, $f_2(x) = g(x) + Kf_1(x), \ldots, f_n(x) = g(x) + Kf_{n-1}(x), \ldots$. If $\{f_n(x)\}$ is convergent to, say, $u(x)$ for $\|\cdot\|_\infty$, then

$$u(x) = \lim f_n(x) = g(x) + \lim Kf_{n-1}(x) = g(x) + Ku(x);$$

i.e., $u(x)$ is a solution to (∗). So in order that (∗) have a solution it is sufficient that the sequence $\{f_n(x)\}$ converge for $\|\cdot\|_\infty$.

For any $v(x)$ in $\mathscr{C}[0, 1]$ let $K^0 v(x) = v(x)$, $K^1 v(x) = Kv(x)$, and, for $n \geq 2$, let $K^n v(x) = K[K^{n-1} v(x)]$. We have

$$
\begin{aligned}
f_0(x) &= 0, \\
f_1(x) &= g(x) + Kf_0(x) = g(x), \\
f_2(x) &= g(x) + Kf_1(x) = g(x) + Kg(x), \\
&\vdots \\
f_n(x) &= g(x) + Kg(x) + K^2 g(x) + \cdots + K^{n-1} g(x), \\
&\vdots
\end{aligned}
$$

Hence if $n > m$, then

$$f_n(x) - f_m(x) = K^m g(x) + K^{m+1} g(x) + \cdots + K^{n-1} g(x).$$

Thus a sufficient condition for the sequence $\{f_n\}$ to converge in $(\mathscr{C}[0, 1], \|\cdot\|_\infty)$ is that the series $\sum_{n=0}^\infty \|K^n g(x)\|$ be convergent.

EXERCISES 1

∗1. Let $(E, \|\cdot\|)$ be a normed space over K and let $\mathscr{L}(E)$ be the space of all continuous, linear operators on E.

(a) For each u in $\mathscr{L}(E)$ show that

$$\sup\{\|u(x)\| \mid \|x\| \leq 1\}$$

$$= \sup\{\|u(x)\| \mid \|x\| = 1\}$$

$$= \sup\{\|u(x)\| \, \|x\|^{-1} \mid x \neq 0\}.$$

(b) For each u in $\mathscr{L}(E)$ define $\|u\|$ to be the number

$\sup\{\|u(x)\| \mid \|x\| \le 1\}$ (see Exercises 1.1, problem 2b). Prove that $\|\cdot\|$ is a norm for $\mathscr{L}(E)$.

2. Let $(B, \|\cdot\|)$ be a Banach space, let $\mathscr{L}(B)$ be the space of all continuous, linear operators on B, and let $\mathscr{L}_c(B)$ be the set of all compact operators on B.
   (a) Prove the statement made just after Definition 1.
   (b) Show that $\mathscr{L}_c(B)$ is a linear subspace of $\mathscr{L}(B)$.
   (c) If $T \in \mathscr{L}(B)$ and $K \in \mathscr{L}_c(B)$ show that both $T \circ K$ and $K \circ T$ are in $\mathscr{L}_c(B)$.
   *(d) If K is in $\mathscr{L}_c(B)$ and n is any positive integer show that there is an operator S_n in $\mathscr{L}_c(B)$ such that $(I - K)^n = I - S_n$. Hint: Use the binomial theorem.
   (e) If $T \in \mathscr{L}(B)$ and if the range of T is a finite dimensional subspace of B show that $T \in \mathscr{L}_c(B)$.

3. Referring to problems 2 and 1b above show that $\mathscr{L}_c(B)$ is a closed, linear subspace of $\mathscr{L}(B)$.

4. Let $K(x, t)$ be continuous for $0 \le t \le x \le 1$, let $g(x) \in \mathscr{C}[0, 1]$ be given and consider the equation

$$(**) \qquad u(x) = g(x) + \int_0^x K(x, t)u(t)\, dt.$$

This is a Volterra integral equation of the second kind.
   (a) For each $f \in \mathscr{C}[0, 1]$ define $Kf(x)$ to be $\int_0^x K(x, t)f(t)\, dt$. Let K be the operator on $\mathscr{C}[0, 1]$ that takes each $f(x)$ to $Kf(x)$. Show that K is a compact operator on $(\mathscr{C}[0, 1], \|\cdot\|_\infty)$.
   (b) Show that the equation $(**)$ has a solution in $\mathscr{C}[0, 1]$ for any given function $g(x) \in \mathscr{C}[0, 1]$. Hint: If

$$M = \max\{|K(x, t)| \mid 0 \le t \le x \le 1\}$$

then $\sum_{n=0}^\infty \|K^n g\|_\infty \le e^M \|g\|_\infty$.

2. Riesz Theory and Complementary Subspaces

Here we shall continue with our study of compact operators. Throughout this section $(B, \|\cdot\|)$ will denote a Banach space, I the identity operator on B and K a compact operator on B. For each

$n = 0, 1, 2, \ldots$ let N_n be the null space of $(I - K)^n$; recall that $(I - K)^0 = I$, $(I - K)^n = (I - K)(I - K)^{n-1}$ for $n \geq 1$. Clearly $\{0\} = N_0 \subset N_1 \subset N_2 \subset \cdots \subset N_n \subset N_{n+1} \subset \cdots$.

Theorem 1. For each n the space N_n is finite dimensional. Furthermore, there is an integer p such that $N_n \neq N_{n+1}$ for $n = 0, 1, 2, \ldots, p - 1$ but $N_n = N_{n+1}$ for $n \geq p$.

Proof. We have already proved that N_1 is finite dimensional (Section 1, Theorem 1). But since $(I - K)^n = I - S_n$ for some compact operator S_n (Exercises 1, problem 2d) the finite dimensionality of N_n also follows from this theorem.

Suppose that $N_n \neq N_{n+1}$ for $n = 0, 1, 2, \ldots$. For each n choose x_n in N_{n+1} such that $\|x_n\| = 1$ and $\inf\{\|x_n - y\| \mid y \text{ in } N_n\} \geq \frac{1}{2}$. Apply K to this sequence. If $s > r$ then

$$\|Kx_s - Kx_r\| = \|x_s - \{(I - K)x_s + x_r - (I - K)x_r\}\| \geq \frac{1}{2}.$$

But this contradicts the fact that K is a compact operator and we must conclude that $N_n = N_{n+1}$ for some n. Let p be the first integer for which this is so. We shall now show that $N_n = N_{n-1}$ for all $n > p$. If x is in N_n then $(I - K)^{p+1}[(I - K)^{n-p-1}x] = (I - K)^n x = 0$. So the element $(I - K)^{n-p-1}x$ is in N_{p+1}. But $N_{p+1} = N_p$ and so $(I - K)^{n-1}x = (I - K)^p[(I - K)^{n-p-1}x] = 0$. This says that x, an arbitrary element of N_n, is in N_{n-1} provided $n > p$.

Now let R_n be the range of the operator $(I - K)^n$ for $n = 0, 1, 2, \ldots$. Clearly $B = R_0 \supset R_1 \supset R_2 \supset \cdots \supset R_n \supset R_{n+1} \supset \cdots$.

Theorem 2. Each of the spaces R_n is a closed, linear subspace of B. Furthermore, there is an integer q such that $R_n \neq R_{n+1}$ for $n = 0, 1, 2, \ldots, q - 1$ but $R_n = R_{n+1}$ for $n \geq q$.

Proof. By Section 1, Theorem 2, we know that R_1 is closed. Since $(I - K)^n = I - S_n$ for some compact operator S_n (Exercises 1, problem 2d), the fact that each R_n is closed also follows from this theorem.

The argument given in the second paragraph of the proof of Theorem 1 can be used here to show that there is some integer n such that $R_n = R_{n+1}$. Let q be the first such integer. Since $R_{q+1} = R_q$ and $(I - K)R_q = R_{q+1}$ we have $R_{q+2} = (I - K)R_{q+1} = (I - K)R_q$, i.e., $R_{q+2} = R_{q+1}$. The fact that $R_n = R_{n+1}$ for all $n \geq q$ can now be proved by induction.

One might suspect that for a fixed, compact operator K the integers p and q are equal. That this is the case is a corollary of the following theorem.

Theorem 3.   Let q be the integer whose existence was proved in Theorem 2. Then

(a) $N_q \cap R_q = \{0\}$,
(b) $N_q + R_q = \{x + y \mid x \text{ in } N_q, y \text{ in } R_q\} = B$,
(c) $K(N_q) \subset N_q$,
(d) $K(R_q) \subset R_q$.

Furthermore, the restriction of the operator $I - K$ to R_q has a continuous linear inverse.

Proof.   (a) We shall actually prove that $N_m \cap R_q = \{0\}$ for any integer m. Suppose that z is in the intersection. For each $n \geq q$ we must have a point z_n in B such that $z = (I - K)^n z_n$ because $R_n = R_q$ for $n \geq q$. If we assume that $z \neq 0$ then $z_n \notin N_n$ for every n. But clearly z_n does belong to N_{m+n} because z belongs to N_m. However, if n is large enough, $N_{m+n} = N_n$ and we have reached a contradiction.
  (b) If $z \in B$ there is a y in B such that $(I - K)^q z = (I - K)^{2q} y$. Hence $(I - K)^q[z - (I - K)^q y] = 0$ and we have shown that $z - (I - K)^q y$ is in N_q. But then z, which is equal to $[z - (I - K)^q y] + (I - K)^q y$, is in $N_q + R_q$, and so the sum of these spaces must be all of B.
  (c) Note that $K(N_0) \subset N_0$ and $(I - K)N_k \subset N_{k-1} \subset N_k$ for all $k \geq 1$. Hence

$$K(N_k) = [I - (I - K)]N_k \subset N_k + (I - K)N_k \subset N_k + N_k = N_k.$$

  (d) Note that $R_{q+1} = (I - K)R_q = R_q$ and argue as in (c).
  Finally, let $I - K \mid R_q$ denote the restriction of $I - K$ to R_q. It is clear, by (c), that $I - K$ is a continuous, linear operator on R_q. The null space of $I - K \mid R_q$ is $N_1 \cap R_q = \{0\}$ by (a). Thus $I - K \mid R_q$ is one-to-one and hence has a linear inverse. The fact that this inverse is continuous follows from Lemma 1.1.

Corollary 1.   The integers p and q defined in Theorems 1 and 2 are equal.

Proof.   Let x be any point of N_{q+1} and find $y \in N_q$, $z \in R_q$ such that $x = y + z$. Clearly

$$(I - K)^{q+1}z = (I - K)^{q+1}(x - y) = 0.$$

But $z \in R_q$, $I - K \mid R_q$ has an inverse, and $(I - K)R_q \subset R_q$. It follows that $z = 0$ and hence that $x \in N_q$. So $N_{q+1} = N_q$ and we must conclude that $q \geq p$ because p is the first integer for which these spaces coincide. Now $(I - K)^p N_p$ is the zero subspace. Hence

$$R_p = (I - K)^p B$$
$$= (I - K)^p R_q + (I - K)^p N_q = (I - K)^p R_q = R_{q+p} = R_q,$$

and so $p \geq q$.

Theorem 3 is very useful for more detailed studies of the properties of compact operators [3]. It also leads to some interesting questions about Banach spaces. The spaces N_q and R_q are both closed, their intersection contains only the zero vector, and their sum is all of B. Given a closed linear subspace G of a Banach space B is it always possible to find a closed linear subspace H of B such that $G \cap H = \{0\}$ and $G + H = B$? Are such pairs of closed linear subspaces always related to continuous, linear operators on B and, if so, what is the connection between the operator and the spaces? These questions will take us very far in our study of Banach spaces. To begin discussing them we need some terminology.

Definition 1. Let X be a vector space over K. Two linear subspaces Y, Z of X are said to be supplementary subspaces of X (we also say that each is a supplement of the other in X) if their intersection contains only the zero vector and their sum $Y + Z = \{y + z \mid y$ in Y, z in $Z\}$, is all of X.

A linear map P from X into itself is said to be a projection operator on X if $P[Px] = Px$ for every x in X.

Given a projection operator P on X the subspaces $Y = \{x$ in $X \mid Px = 0\}$ and $Z = \{Px \mid x$ is in $X\}$ are supplementary subspaces of X. Note that $I - P$ is also a projection operator on X, that the range of this operator is equal to the null space of P, and that the null space of this operator is equal to the range of P. Thus if P is a continuous projection operator on a Banach space B then both the null space of P and the range of P are closed.

The even and odd functions illustrate these ideas. Consider the Banach space $(\mathscr{C}[-1, 1], \|\cdot\|_\infty)$. We mean, of course, the space of continuous functions on $[-1, 1]$ with $\|f\|_\infty = \max\{|f(x)\| - 1 \le x \le 1\}$. For any point f in this space define $Pf(x)$ to be $2^{-1}\{f(x) + f(-x)\}$. The operator P that takes $f(x)$ in $\mathscr{C}[-1, 1]$ to $Pf(x)$ is clearly a continuous projection operator. Hence the range of P, $\{f \mid f(-x) = f(x)$ for all x in $[-1, 1]\}$, and the null space of P, $\{f \mid f(-x) = -f(x)$ for all x in $[-1, 1]\}$, are supplementary subspaces of $\mathscr{C}[-1, 1]$, and each of them is closed.

Definition 2. Let B be a Banach space and let G, H be two supplementary subspaces of B. If both of these spaces is closed in B, then we shall say that they are complementary subspaces of B (we shall also say that each is a complement of the other in B).

EXERCISES 2

1. Let f be a continuous, linear functional on a Banach space B (Section 1.3, Definition 2). Suppose that $x_0 \in B$ and $f(x_0) = 1$. Define Tx, for each x in B, to be $f(x)x_0$.
    (a) Show that T is a compact operator on B.
    (b) Identify the spaces N_1 and R_1 of Theorems 1 and 2. What is the integer p in this case?

2. Let T be a compact operator on a Banach space B. Show that the range of T is a separable, linear subspace of B (Section 1.3, Definition 4).

3. Let X be a vector space over K.
    (a) Show that every linear subspace of X has a supplement in X.
    (b) Let Y, Z be two linear subspaces of X. Show that Z is a supplement for Y in X iff there is a projection operator on X whose range is Y and whose null space is Z.

4. Give an example of a pair of complementary subspaces in $(l_\infty, \|\cdot\|_\infty)$ and $(l_1, \|\cdot\|_1)$.

*5. Let B be a Banach space and let G be a closed linear subspace of B. Let H be a subspace of B that is a supplement for G in B. It is instructive to try to prove that H is closed or that the closure of H

is a complement for G. We shall see (Chapter 3) that neither of these need be true.

6. Let X be a vector space over K. We shall say that a linear subspace Y of X has finite codimension in X if the quotient space X/Y is finite dimensional. The dimension of the quotient space is taken to be the codimension of Y in X.

(a) If Y has codimension n in X show that there is a linearly independent set $\phi_1, \phi_2, \ldots, \phi_n$ in $X^\#$ such that, if $N(\phi_j)$ is the null space of ϕ_j for each j, $Y = \bigcap_{j=1}^n N(\phi_j)$.

*(b) Let $\phi_1, \phi_2, \ldots, \phi_p$ be a (finite) linearly independent subset of $X^\#$. Suppose that $\phi_0 \in X^\#$ and that $N(\phi_0) \supset \bigcap_{j=1}^p N(\phi_j)$. Show that $\phi_0, \phi_1, \ldots, \phi_p$ is a linearly dependent set. Hint: Use induction on p. If $p = 1$ the result is already known (Exercise 3, problem 4b). Since ϕ_2 is a linear functional on $N(\phi_1)$, $N(\phi_1) \cap N(\phi_2)$ has codimension two in X. Similarly, if $N = \bigcap_{j=1}^p N(\phi_j)$, X/N has a dimension $\leq p$. Since the dimension of X/N is equal to that of $(X/N)^\#$, these spaces have dimension p.

(c) Let $(E, \|\cdot\|)$ be a normed space and let G be a closed, linear subspace of E that has finite codimension in E. Prove that G has a complement in E.

3. The Open-Mapping Theorem

Given a continuous projection on a Banach space B it is clear (see the next to last paragraph in Section 2) that the range and the null space of this projection are complementary subspaces of B. Suppose now that a pair of complementary subspaces, say G and H, of B are given. We may ask whether there exists a continuous projection on B whose range is G and whose null space is H. Since $G + H = B$ and $G \cap H = \{0\}$ we may define $P(g + h)$ to be g for all g in G, h in H. Then P is certainly a projection on B, the range of P is G, and the null space of P is H. But is P continuous? Let us look more closely at this operator.

We can define a map ϕ from the product space $G \times H$ onto B by letting $\phi[(g, h)] = g + h$. This map is linear and one-to-one so ϕ^{-1}

exists, and it is a linear map. We can also define a map π from $G \times H$ onto G by letting $\pi[(g, h)] = g$. Notice that $P = \pi \circ \phi^{-1}$.

For each pair (g, h) in $G \times H$ let $\|(g, h)\|$ be the maximum of the numbers $\|g\|$, $\|h\|$. We leave it to the reader to show that we have actually defined a norm on $G \times H$ and that, with this norm, $G \times H$ is a Banach space (problem 1 below). Since $\|\pi[(g, h)]\| = \|g\| \le \|(g, h)\|$ and $\|\phi[(g, h)]\| \le 2\|(g, h)\|$, both of these maps are continuous. But in order to prove that P is continuous we have to show that ϕ^{-1} is continuous. Summing up what we know about ϕ we are led to the following question: Does a linear, one-to-one, continuous map from one Banach space onto a second Banach space have a continuous inverse? This difficult question will occupy us for the remainder of this section. Our first two results are rather technical.

Lemma 1. A normed space $(E, \|\cdot\|)$ is a Banach space iff it satisfies the following condition: For any sequence $\{x_i\}$ of points of E the sequence $\{\sum_{i=1}^{n} x_i \,|\, n = 1, 2, \ldots\}$ converges to a point of E whenever the series $\sum_{i=1}^{\infty} \|x_i\|$ is convergent.

Proof. Assume that $(E, \|\cdot\|)$ is a Banach space, let $\{x_i\}$ be a sequence of points of E, and assume that $\sum_{k=1}^{\infty} \|x_i\|$ is convergent. For any two positive integers m, n, with $m > n$, we have

$$\left\| \sum_{i=1}^{m} x_i - \sum_{i=1}^{n} x_i \right\| \le \sum_{i=n+1}^{m} \|x_i\|.$$

Since this last sum tends to zero as m and n tend to infinity, $\{\sum_{i=1}^{n} x_i \,|\, n = 1, 2, \ldots\}$ is a Cauchy sequence in $(E, \|\cdot\|)$.

Now assume that the normed space $(E, \|\cdot\|)$ satisfies our condition. Let $\{y_n\}$ be a Cauchy sequence in this space. We may choose a subsequence $\{z_n\}$ of $\{y_n\}$ such that $\|z_i - z_j\| < 2^{-i}$ for all $j \ge i$. Let $x_1 = z_1$ and, for $i > 1$, let $x_i = z_i - z_{i-1}$. Clearly $\sum_{i=1}^{n} x_i = z_n$ and $\sum_{i=1}^{n} \|x_i\| \le \|z_1\| + 1$ for all n. It follows that $\lim \sum_{i=1}^{n} x_i = \lim z_n$ exists in E. But since $\{z_n\}$ is a subsequence of $\{y_n\}$ and since $\{y_n\}$ is a Cauchy sequence, $\{y_n\}$ must converge to a point of E.

Lemma 2. Let $(B_1, \|\cdot\|_1)$ and $(B_2, \|\cdot\|_2)$ be two Banach spaces over the same field and suppose that T is a continuous linear map from B_1 onto B_2. If S_0 is $\{x \in B_1 \,|\, \|x\|_1 < 1\}$ then for some $r > 0$ the set $T(S_0) = \{Tx \,|\, x \text{ in } S_0\}$ contains $\{y \in B_2 \,|\, \|y\|_2 < r\}$.

Proof. For each $n = 0, 1, 2, \ldots$ let $S_n = \{x \in B_1 \,|\, \|x\|_1 < 2^{-n}\}$.

Since $B_1 = \bigcup_{k=1}^{\infty} kS_1$ and T is onto we must have $B_2 = \bigcup_{k=1}^{\infty} kT(S_1)$. By the Baire category theorem [19, Corollary to Theorem 15, p. 139] there is an integer k for which the closure of the set $kT(S_1)$ (denote this by $cl[kT(S_1)]$) has nonempty interior. Thus for some point z' in B_2 and some positive real number η', $cl[kT(S_1)] \supset \{y \text{ in } B_2 \,|\, \|y - z'\|_2 < \eta'\}$. Now the map that takes each y in B_2 onto ky is a homeomorphism from $(B_2, \|\cdot\|_2)$ onto itself. Hence $cl[T(S_1)]$ contains a ball; i.e., there is a point z in B_2 and a positive number η such that $cl[T(S_1)] \supset \{y \text{ in } B_2 \,|\, \|y - z\|_2 < \eta\}$. Clearly $cl[T(S_1)] \supset \{y + z \,|\, \|y\|_2 < \eta\}$ and so $cl[T(S_1)] - z$ contains $\{y \text{ in } B_2 \,|\, \|y\|_2 < \eta\}$. But since z is in $cl[T(S_1)]$,

$$cl[T(S_1)] - z \subset cl[T(S_1)] - cl[T(S_1)] \subset 2cl[T(S_1)]$$
$$= cl[2T(S_1)] = cl[T(S_0)].$$

We have shown that $cl[T(S_0)]$ contains an open ball, centered at zero in B_2, having radius η. Clearly $cl[T(S_n)]$ will contain an open ball, centered at zero in B_2, having radius $\eta 2^{-n}$, $n = 1, 2, \ldots$.

Now let y be a point of B_2, $\|y\|_2 < \eta 2^{-1}$. Since y is in the closure of $T(S_1)$ we can find a point x_1 in S_1 such that $\|y - Tx_1\|_2 < \eta 2^{-2}$. Then the point $y - Tx_1$ is in the closure of the set $T(S_2)$, and so there is a point x_2 in S_2 such that $\|(y - Tx_1) - Tx_2\|_2 < \eta 2^{-3}$. We can continue this selection process. After $x_1, x_2, \ldots, x_{n-1}$ have been chosen we choose x_n in S_n such that

$$\left\| \left(y - \sum_{i=1}^{n-1} Tx_i \right) - Tx_n \right\|_2 < \eta 2^{-(n+1)}.$$

Consider the sequence $\{x_i\}$ of B_1 that we obtain. We have $\sum_{i=1}^{\infty} \|x_i\| < \sum_{i=1}^{\infty} 2^{-i} = 1$ and so, by Lemma 1, the sequence $\{\sum_{i=1}^{n} x_i \,|\, n = 1, 2, \ldots\}$ converges to a point of B_1 for $\|\cdot\|_1$. Denote the limit of this last sequence by $\sum_{i=1}^{\infty} x_i$, observe that this point is in S_0, and that $T(\sum_{i=1}^{\infty} x_i) = \sum_{i=1}^{\infty} T(x_i) = y$. Hence y is in $T(S_0)$. It follows that $\{y \text{ in } B_2 \,|\, \|y\|_2 < \eta 2^{-1}\}$ is contained in $T(S_0)$.

Remark. It is worthwhile observing that the proof of Lemma 2 shows the following: The image under a continuous, linear map of one Banach space in a second Banach space is either all of the second Banach space or it is a set of the first category in the latter space.

We shall now prove that the question raised in the beginning of this section (concerning continuity of the inverse) has an affirmative answer. This fact is a special case of the next result.

Theorem 1 (Open-Mapping Theorem).   Let T be a continuous, linear map from one Banach space onto a second Banach space. Then T maps any open subset of its domain onto an open subset of its range.

Proof.   Let B_1, B_2 be the domain and the range, respectively, of T. These two spaces are, by hypothesis, Banach spaces. If θ is an open subset of B_1, if y is a point of $T(\theta)$, and if x is a point of θ such that $Tx = y$, then there is an open ball $\mathscr{B}$, centered at x, such that $\mathscr{B} \subset \theta$ for θ is an open set. Clearly $\mathscr{B} - x$ is an open ball centered at zero in B_1. By Lemma 2 the set $T(\mathscr{B} - x)$ contains an open ball $\mathscr{B}'$, centered at zero, in B_2. But $T(\mathscr{B}) - y = T[\mathscr{B} - x] \supset \mathscr{B}'$. This says that $T(\mathscr{B})$, and hence $T(\theta)$, contains $y + \mathscr{B}'$, i.e., $T(\theta)$ is a neighborhood of y.

Referring again to the discussion given at the beginning of this section we have:

Corollary 1.   Let B be a Banach space and let G be a closed, linear subspace of B. Then G has a complement in B iff there is a continuous projection operator on B whose range is G.

EXERCISES 3

*1.   Let $(E_1, \|\cdot\|_1)$ and $(E_2, \|\cdot\|_2)$ be two normed spaces over the same field. For each pair (x_1, x_2) in $E_1 \times E_2$ define $\|(x_1, x_2)\|_p$ to be the maximum of the numbers $\|x_1\|_1$ and $\|x_2\|_2$.
   (a)   Prove that $\|\cdot\|_p$ is a norm for $E_1 \times E_2$. This is called the product space norm for $E_1 \times E_2$. Whenever we work with a product of two normed spaces we shall always assume that it has the product space norm.
   (b)   Let π_1 be defined on $E_1 \times E_2$ by $\pi_1[(x_1, x_2)] = x_1$. Show that π_1 is a continuous map from $E_1 \times E_2$ onto E_2.
   (c)   If both $(E_1, \|\cdot\|_1)$ and $(E_2, \|\cdot\|_2)$ are Banach spaces prove that $(E_1 \times E_2, \|\cdot\|_p)$ is a Banach space.

2.   Let $(E, \|\cdot\|)$ be a normed space and let α be a nonzero scalar. Define a map A from E into itself by letting $Ax = \alpha x$ for all x in E. Show that A is a homeomorphism from $(E, \|\cdot\|)$ onto itself.

3.   For this problem we must recall the Banach spaces

$(L_p[0, 1], \|\cdot\|_p)$ from real analysis. These spaces are defined in Chapter 6. A detailed treatment of them can be found in [10] or [21]. Fix $p > 1$.

(a)   Prove $L_p[0, 1] \subset L_1[0, 1]$, but $L_p[0, 1] \neq L_1[0, 1]$.

(b)   Prove that the inclusion map from $(L_p[0, 1], \|\cdot\|_p)$ into $(L_1[0, 1], \|\cdot\|_1)$ is continuous.

(c)   Conclude (see the remark after Lemma 2) that $L_p[0, 1]$ is a set of the first category in $(L_1[0, 1], \|\cdot\|_1)$.

4.   Let $(B, \|\cdot\|)$ be a Banach space and let $\|\cdot\|_1$ be a second norm on B. Suppose that there is a constant M such that $\|x\| \leq M\|x\|_1$ for all x in B. If $(B, \|\cdot\|_1)$ is also a Banach space prove that the norms $\|\cdot\|$ and $\|\cdot\|_1$ are equivalent.

4.   Quotient Spaces of l_1

We have gotten away from the question raised in Section 2. Recall that we asked whether every closed, linear subspace of a Banach space B has a complement in B. The results of this section will be useful in settling this question. Here we are going to use the open-mapping theorem to prove an interesting representation theorem for separable (Section 1.3, Definition 4) Banach spaces.

Let $(B, \|\cdot\|)$ be a separable Banach space and recall the space of sequences $(l_1, \|\cdot\|_1)$ defined in Exercises 1.2, problem 5. If $\{x_j\}$ is a fixed, countable subset of $\mathscr{B}$, the unit ball of $(B, \|\cdot\|)$, which is dense in $\mathscr{B}$, then we can define a linear map T from l_1 into $\mathscr{B}$ as follows: For any point $\{t_j\}$ of l_1 let $T[\{t_j\}] = \sum_{j=1}^{\infty} t_j x_j$. Since

$$\|T[\{t_j\}]\| \leq \sum_{j=1}^{\infty} |t_j| \|x_j\| \leq \sum_{j=1}^{\infty} |t_j| = \|\{t_j\}\|_1,$$

T is a continuous linear map from $(l_1, \|\cdot\|_1)$ into $(B, \|\cdot\|)$. Let us show that T is onto.

For any point x in $\mathscr{B}$ choose x_{n_1} in $\{x_j\}$ such that $\|x - x_{n_1}\| < \frac{1}{2}$. The set $\{\frac{1}{2}x_j \mid j \neq n_1\}$ is dense in $\frac{1}{2}\mathscr{B}$, and so there is a point x_{n_2} in $\{x_j\}$ such that $\|(x - x_{n_1}) - \frac{1}{2}x_{n_2}\| < 2^{-2}$. We can even assume that $n_2 > n_1$. Continue selecting points in this way. After $x_{n_1}, x_{n_2}, \ldots, x_{n_k}$ have been chosen, choose $x_{n_{k+1}}$ satisfying $\|x - \sum_{i=1}^{k+1} 2^{-i+1}x_{n_i}\| < 2^{-k-1}$ and with $n_{k+1} > n_k$. Define $\{t_n\} \in l_1$ as follows: $t_n = 0$ if $n \neq n_k$ for all k;

$t_n = 2^{-k+1}$ if $n = n_k$. Clearly $T[\{t_n\}] = x$ and so $\mathcal{B}$ is contained in the range of T.

Let G be the null space of T. Then, of course, $G \subset l_1$ and the linear map T induces an isomorphism from l_1/G onto B. Call this isomorphism T^*. If l_1/G has a norm, say $\|\cdot\|_q$, if $(l_1/G, \|\cdot\|_q)$ is a Banach space, and if T^* is continuous from $(l_1/G, \|\cdot\|_q)$ onto $(B, \|\cdot\|)$, then by the open-mapping theorem T^* is a topological isomorphism. In other words, if we can define a norm $\|\cdot\|_q$ on l_1/G with the properties just stated, then we shall have proved:

Theorem 1. Every separable Banach space is topologically isomorphic to some quotient space of $(l_1, \|\cdot\|_1)$.

Let $(E, \|\cdot\|)$ be a normed space and let G be a linear subspace of E. For each x in E let $\dot{x}$ denote the element of E/G that contains x. Define a function $\|\cdot\|_q$ on E/G as follows: $\|\dot{x}\|_q = \inf\{\|x + g\| \,|\, g \text{ in } G\}$ for each $\dot{x}$ in E/G. This is clearly a nonnegative function on E/G, and it is easy to see that $\|\dot{x} + \dot{y}\|_q \le \|\dot{x}\|_q + \|\dot{y}\|_q$ and $\|\alpha\dot{x}\|_q = |\alpha| \, \|\dot{x}\|_q$ for all x, y in E/G and all α in K.

Lemma 1. The function $\|\cdot\|_q$ is a norm for E/G iff G is closed.

Proof. Suppose that G is a closed, linear subspace of E. Let $\dot{x}$ be an element of E/G such that $\|\dot{x}\|_q = 0$, and let x be a fixed element of $\dot{x}$. For any positive integer n there is a point g_n in G such that $\|x + g_n\| < 1/n$. But then the sequence $\{-g_n\}$ of points of G converges to x. Since G is closed it follows that $x \in G$ and so $\dot{x} = 0$.

Now assume that G is not closed and let x be a point that is not in G but is in the closure of G. If $\{g_n\}$ is a sequence of points of G that converges to x then $\|\dot{x}\|_q \le \inf\{\|x - g_n\| \,|\, n = 1, 2, \ldots\}$. Since the infimum is zero, $\|\dot{x}\|_q = 0$. But $\dot{x} \ne 0$ and so $\|\cdot\|_q$ could not be a norm on E/G.

Definition 1. Let $(E, \|\cdot\|)$ be a normed space and let G be a closed, linear subspace of E. The norm on E/G defined just before the statement of Lemma 1 is called the quotient norm on E/G.

Whenever we work with the quotient space of a normed space and one of its closed, linear subspaces, we shall always assume, without explicit mention, that it has the quotient space norm. We shall also omit the subscript q on the quotient norm.

Theorem 2.   Let $(B, \|\cdot\|)$ be a Banach space and let G be a closed, linear subspace of B. Then B/G is a Banach space.

Proof.   Let $\{\dot{x}_n\}$ be a Cauchy sequence in B/G. Then there is a subsequence $\{\dot{x}_{n_k}\}$ of $\{\dot{x}_n\}$ such that $\|\dot{x}_{n_k} - \dot{x}_{n_m}\| < 2^{-k}$ for all $m \geq k$; in particular $\|\dot{x}_{n_{k+1}} - \dot{x}_{n_k}\| < 2^{-k}$ for every k. It follows, from Definition 1, that there are elements x_{n_1} of $\dot{x}_{n_1}$, x_{n_2} of $\dot{x}_{n_2}$, ... such that $\|x_{n_{k+1}} - x_{n_k}\| < 2^{-k}$ for all k. Now if $l > k$ then

$$\|x_{n_l} - x_{n_k}\| \leq \|x_{n_l} - x_{n_{l-1}}\| + \cdots + \|x_{n_{k+1}} - x_{n_k}\|$$

$$< 2^{-l+1} + \cdots + 2^{-k} < 2^{-k+1}.$$

Hence $\{x_{n_k}\}$ is a Cauchy sequence in the Banach space $(B, \|\cdot\|)$. Let $x \in B$ be the limit of this sequence. Since $\|\dot{x}_{n_k} - \dot{x}\| \leq \|x_{n_k} - x\|$, $\{\dot{x}_{n_k}\}$ converges to $\dot{x}$ for the norm of B/G. But then $\{\dot{x}_n\}$ must converge to $\dot{x}$ for this norm.

EXERCISES 4

1. Let $(E, \|\cdot\|)$ be a normed space and let G be a closed linear subspace of E. For each x in E let $N(x)$ be the unique element of E/G that contains x. Let N be the map that takes each x in E to $N(x) \in E/G$. Show that N is continuous.

2. Let $(E, \|\cdot\|)$ be a normed space and let $(B, \|\|\cdot\|\|)$ be a Banach space. Suppose that T is a continuous, linear map from E onto B with null space G. We may define a map T^* from E/G onto B as follows: For each $\dot{x}$ in E/G choose x in $\dot{x}$ arbitrarily, and let $T^*\dot{x} = Tx$.
    (a) Show that T^* is a continuous, linear map from E/G onto $(B, \|\|\cdot\|\|)$.
    (b) If $(E, \|\cdot\|)$ happens to be a Banach space show that T^* is a topological isomorphism.

*3. Let $(B, \|\cdot\|)$ be a Banach space and let G be a closed, linear subspace of B. Show that any two complements of G in B are topologically isomorphic.

5. The Closed Graph Theorem

The closed graph theorem is a particularly useful relative of the open-mapping theorem.

Definition 1. Let T be a linear map from the normed space $(E_1, \|\cdot\|_1)$ into the normed space $(E_2, \|\cdot\|_2)$. The set $\{(x, Tx) \mid x \text{ in } E_1\}$, which we shall denote by $\mathscr{G}(T)$, is called the graph of T.

We leave it to the reader to show that $\mathscr{G}(T)$ is a linear subspace of $E_1 \times E_2$ and that, if T is continuous, $\mathscr{G}(T)$ is closed in the product space. The remarkable fact is that, for Banach spaces, the converse is true.

Theorem 1 (Closed Graph Theorem). A linear map from one Banach space into a second Banach space is continuous iff its graph is closed in the product of the two spaces.

Proof. Let T be a linear map from the Banach space $(B_1, \|\cdot\|_1)$ into the Banach space $(B_2, \|\cdot\|_2)$. If $\mathscr{G}(T)$ is closed in $B_1 \times B_2$ then, since $B_1 \times B_2$ is a Banach space (Exercises 3, problem 1c), $\mathscr{G}(T)$, with the subspace norm, is itself a Banach space. For each (x, y) in $B_1 \times B_2$ let $p(x, y) = x$, $q(x, y) = y$, and let π be the restriction of p to $\mathscr{G}(T)$. By the open-mapping theorem π^{-1} is a continuous, linear map from $(B_1, \|\cdot\|_1)$ onto $\mathscr{G}(T)$. Hence the continuity of the map T follows from the equation $T = q \circ \pi^{-1}$.

Corollary 1. Let $(B, \|\cdot\|)$ be a Banach space and let G, H be two closed, linear subspaces of B such that $G \cap H = \{0\}$. Then $G + H$ is closed iff there is a constant α such that $\|x\| \leq \alpha \|x + y\|$ for all x in G and all y in H.

Proof. Assume that $G + H$ is closed in $(B, \|\cdot\|)$ and define a linear map T from $G + H$ onto G by setting $T(x + y) = x$ for all $x + y$ in

$G + H$. We shall show that $\mathscr{G}(T)$ is a closed subset of $(G + H) \times G$. Let $\{(w_n, Tw_n)\}$ be a sequence in $\mathscr{G}(T)$ that converges to some point of this product space. Then $\{w_n\}$ and $\{Tw_n\}$ must both converge. Writing $w_n = x_n + y_n$, where $x_n \in G$ and $y_n \in H$ for all n, we see that $\{w_n\}$ and $\{x_n\}$ must both converge. It follows that $\{y_n\}$ is convergent. Since G, H, and $G + H$ are closed, these sequences converge to points w_0, x_0, y_0 in $G + H$, G, and H, respectively. But w_0 is clearly $x_0 + y_0$ and so $Tw_0 = x_0$, i.e., $\mathscr{G}(T)$ is a closed set. By Theorem 1, T is continuous and hence $\|x\| \leq \|T\|\|x + y\|$ for all x in G, y in H (Exercises 1.1, problem 2b and Exercises 2.1, problem 1b).

The proof that our condition is sufficient to insure that $G + H$ is closed is left to the reader.

The following example shows that the open-mapping theorem need not be true for normed spaces that are not Banach spaces. Consider $(\mathscr{C}[0, 1], \|\cdot\|)$ and let F be the linear subspace of $\mathscr{C}[0, 1]$ consisting of all functions f such that f' exists on $(0, 1)$ and is uniformly continuous on that interval. If $f \in F$ then there is one and only one function that is continuous on $[0, 1]$ and is equal to f' on $(0, 1)$. Denote this function by f' also. Thus, defining $Tf = f'$ for all $f \in F$, we have a linear map from the normed space F into the Banach space $\mathscr{C}[0, 1]$. Since $F \neq \mathscr{C}[0, 1]$ but F does contain every polynomial function, F can not be a Banach space. The set $\{x^n \mid n = 1, 2, 3, \ldots\}$ is bounded (Section 1.2, Definition 2) in F. However, $\{Tx^n \mid n = 1, 2, \ldots\} = \{nx^{n-1} \mid n = 1, 2, \ldots\}$ is not bounded in $(\mathscr{C}[0, 1], \|\cdot\|_\infty)$. It follows (Exercises 1.2, problem 3a) that T can not be continuous. But it is easy to see that the graph of T is closed.

EXERCISES 5

1. Let $(E_1, \|\cdot\|_1)$, $(E_2, \|\cdot\|_2)$ be two normed spaces, let T be a linear map from E_1 into E_2, and let $\mathscr{G}(T)$ be the graph of T.
   (a) Show that $\mathscr{G}(T)$ is a linear subspace of $E_1 \times E_2$.
   (b) If T is continuous prove that $\mathscr{G}(T)$ is closed in $E_1 \times E_2$.
2. Referring to Corollary 1 prove that the condition stated is sufficient to insure that $G + H$ is closed in B.
3. Let T be a linear, one-to-one, continuous map from the Banach

space $(B_1, \|\cdot\|_1)$ onto the Banach space $(B_2, \|\cdot\|_2)$. Use the closed graph theorem to prove that T has a continuous inverse. Hint: The map $(x, y) \rightarrow (y, x)$ is an isomorphism from $\mathscr{G}(T)$ onto $\mathscr{G}(T^{-1})$.

4. Show that the discontinuous, linear map defined at the end of this section has a closed graph.

CHAPTER 3

Linear Functionals

1. Special Subspaces of l_∞ and l_1.
The Dual Space

We have been talking about complementary subspaces of a Banach space for some time now. We know about their connection with continuous, projection operators and we have seen some examples. But the fact is that a closed, linear subspace of a Banach space may not have a complement. Here are some examples:

(a) The Banach space $(l_\infty, \|\cdot\|)$ consists of all bounded sequences $\{x_n\}$ with $\|\{x_n\}\| = \sup\{|x_n| \mid n = 1, 2, \ldots\}$. For each fixed, positive integer k let $f_k(x) = f_k(\{x_n\}) = x_k$ for all x in l_∞. So f_k maps each bounded sequence onto its kth term. Clearly each f_k is a continuous, linear functional on $(l_\infty, \|\cdot\|_\infty)$ (Section 1.3, Definition 2) and if, for x in $l_\infty, f_k(x) = 0$ for all k, then $x = 0$.

The sequences that converge to zero comprise a closed, linear subspace (we called it c_0 (Exercises 1.3, problem 2)) of l_∞. If c_0 had a complement H in l_∞ then there would be a topological isomorphism ϕ from l_∞/c_0 onto H (Exercises 2.4, problem 3). For each k, let $h_k = f_k \circ \phi$. Then $\{h_k\}$ is a countable set of continuous, linear functionals on l_∞/c_0 that has the following property:

(∗)   If x is in l_∞/c_0 and if $h_k(x) = 0$ for all k, then $x = 0$.

We shall show that c_0 does not have a complement in l_∞ by showing that no countable set of continuous, linear functionals on l_∞/c_0 can have property (∗).

Let Z_+ denote the set of all positive integers. We can regard l_∞ as the space of all bounded functions on Z_+. If U is a subset of Z_+ then the function that is one at each point of U and is zero at each point that is not in U is called the characteristic function of U. Any such function is in l_∞. We must now prove some facts about Z_+ and about l_∞/c_0.

(i)   There is an uncountable family $\{U_a \mid a \text{ in } A\}$ of subsets of Z_+ such that each U_a is an infinite set, and $U_a \cap U_b$ is finite for $a \neq b$.

Let ψ be a one-to-one correspondence from the rationals in $(0, 1)$ onto Z_+ and let A be the set of all irrationals in $(0, 1)$. For each a in A let U'_a be the terms of any sequence of rationals in $(0, 1)$ that converges to a. Setting $U_a = \psi(U'_a)$ for each a in A we obtain a family of sets with the required properties.

(ii)   For each a in A let x_a be the element of l_∞/c_0 that contains the characteristic function of the set U_a (we mean, of course, the family described in (i)). Let g be any continuous, linear functional on l_∞/c_0. Then $\{x_a \mid g(x_a) \neq 0\}$ is countable.

We need only show that, for each n in Z_+, $\{x_a \mid |g(x_a)| \geq 1/n\}$ is finite. Choose and fix n and let $x_1, \ldots, x_m$ be in the set under discussion. Let $b_j = \overline{g(x_j)} \, |g(x_j)|^{-1}$ for $j = 1, 2, \ldots, n$ (the bar denotes the complex conjugate) and let $x = \sum_{j=1}^m b_j x_j$. Clearly $x \in l_\infty/c_0$, the norm of x is less than or equal to one, and $g(x) \geq m/n$. Since g is continuous, m must be finite.

If $\{h_j\}$ is any countable set of continuous, linear functionals on l_∞/c_0, then there are only countably many x_a such that $h_j(x_a) \neq 0$ for $j = 1, 2, \ldots$. Since A is uncountable this implies that for some $x_a \neq 0$ every $h_j(x_a) = 0$. Thus c_0 is a closed, linear subspace of $(l_\infty, \|\cdot\|_\infty)$ that has no complement in l_∞.

The first person to prove that c_0 does not have a complement in l_∞ seems to have been Sobczyk [23]. He observed that this is an easy consequence of a result of Phillips [20]. The elegant proof given above is due to Whitely [25].

(b) In order to present our next example we shall need some preliminary results. These are interesting in their own right. If $\{y_n\}$ is in l_∞ and $\{x_n\}$ is any sequence such that $\sum |x_n| < \infty$, then

$$\left|\sum x_n y_n\right| \le \sum |x_n y_n| \le \|\{y_n\}\|_\infty \sum |x_n|.$$

Hence $\sum x_n y_n$ is convergent and we have shown, moreover, that every element of l_∞ defines a continuous, linear functional on $(l_1, \|\cdot\|_1)$.

The sequence $e_1, e_2, \ldots, e_k, \ldots$, where $e_k = (0, 0, \ldots, 0, 1, 0, 0, \ldots)$ (the 1 is in the kth place) is a Schauder basis for $(l_1, \|\cdot\|_1)$ (Exercises 1.3, problem 3). For any continuous, linear functional f on $(l_1, \|\cdot\|_1)$ the sequence $\{f(e_j)\}$ is in l_∞. If x is in l_1 then $\lim \|x - \sum_{j=1}^n a_j e_j\|_1 = 0$ for some unique sequence of constants $a_1, a_2, \ldots$. Clearly $f(x) = \sum a_j f(e_j)$, i.e., the continuous linear functional f is defined by the element $\{f(e_j)\}$ of l_∞. Thus the space of all continuous, linear functionals on $(l_1, \|\cdot\|_1)$ is isomorphic to the vector space l_∞.

Lemma 1. A sequence $\{x_n\}$ of points of l_1 is norm convergent to zero iff $\lim f(x_n) = 0$ for any continuous, linear functional f on l_1.

The proof is rather tricky and so we shall try to give the idea behind it first. Suppose that $\{x_n\}$ is a sequence in l_1. Each x_n is itself a sequence, say $x_n = \{t_i(n) \mid i = 1, 2, \ldots\}$. Assume that $\lim f(x_n) = 0$ for each continuous linear functional f on l_1. Then, since $e_1, e_2, \ldots$ are in l_∞, each $t_i(n)$ converges to zero as n tends to infinity, i.e., $\lim_{n\to\infty} t_i(n) = 0$ for $i = 1$, $2, \ldots$. If $v = \{v_i\}$ is in l_∞, then $v(x_n) = \sum_{i=1}^\infty v_i t_i(n)$ and clearly $\lim_{n\to\infty} v_i t_i(n) = 0$ for $i = 1, 2, \ldots$. Now write

$$v(x_n) = \sum_{i=1}^{n-1} v_i t_i(n) + v_n t_n(n) + \sum_{i=n+1}^\infty v_i t_i(n).$$

Suppose, on the assumption that $\{x_n\}$ does not tend to zero for the l_1 norm, that we can find $\varepsilon > 0$ and $v = \{v_i\}$ in l_∞ with $|v_i| = 1$ for all i such that $\left|\sum_{i=1}^{n-1} v_i t_i(n)\right| < \varepsilon/5$, $|v_n t_n(n)| > 3\varepsilon/5$, and

$$\left|\sum_{i=n+1}^\infty v_i t_i(n)\right| < \varepsilon/5$$

for each n. When $n = 1$ we have $|v_1 t_1(1)| > 3\varepsilon/5$, i.e., we have a "hump" in the first term. We know the hump cannot stay there because $v_1 t_1(n)$ tends to zero as n increases. But when $n = 2$, $|v_2 t_2(2)| > 3\varepsilon/5$, so our hump now appears in the second term. It moved. When $n = 3$ the hump is in the third term, and so on. As n increases the hump "glides" toward infinity. Now clearly, if we can do this,

$$|v(x_n)|$$

$$= \left| \sum_{i=1}^{\infty} v_i t_i(n) \right| \geq |v_n t_n(n)| - \left| \sum_{i=1}^{n-1} v_i t_i(n) \right| - \left| \sum_{i=n+1}^{\infty} v_i t_i(n) \right| \geq \varepsilon/5,$$

so $\lim v(x_n)$ is not zero and we have our contradiction. The proof below is a refinement of this idea.

Proof of Lemma 1. The necessity of our condition is clear. Assume that $\{x_n\}$ satisfies this condition and, for each n, let $x_n = \{t_i(n) \,|\, i = 1, 2, 3, \ldots\}$. If this sequence does not converge to zero for the norm of l_1, then there is an $\varepsilon > 0$ and infinitely many integers $n_j, j = 1, 2, \ldots$, such that $\|x_{n_j}\|_1 = \sum_{i=1}^{\infty} |t_i(n_j)| > \varepsilon$. Choose N_1 so large that

$$\sum_{N_1+1}^{\infty} |t_i(n_1)| \leq \varepsilon/5 \qquad \text{and} \qquad \sum_{i=1}^{N_1} |t_i(n_1)| > 4\varepsilon/5.$$

Next choose complex numbers $v_1, v_2, \ldots, v_{N_1}$ of modulus one so that

$$\sum_{i=1}^{N_1} v_i t_i(n_1) = \sum_{i=1}^{N_1} |t_i(n_1)| > 4\varepsilon/5.$$

Observe that, if v_i is any complex number of modulus one when $i > N_1$, $|\sum v_i t_i(n_1)| > 3\varepsilon/5 > \varepsilon/5$. Now take j_2 so large that n_{j_2} is large enough to imply that $\sum_1^{N_1} |t_i(n_{j_2})| \leq \varepsilon/5$; this is possible because of our assumption on $\{x_n\}$ and the fact that the sequences $e_1, e_2, \ldots$ are in l_∞. Now take $N_2 > N_1$ so that

$$\sum_{N_2+1}^{\infty} |t_i(n_{j_2})| \leq \varepsilon/5 \qquad \text{and} \qquad \sum_1^{N_2} |t_i(n_{j_2})| > 4\varepsilon/5.$$

We can choose complex numbers $v_{N_1+1}, v_{N_1+2}, \ldots, v_{N_2}$ of modulus one so that

$$\sum_{N_1+1}^{N_2} v_i t_i(n_{j_2}) = \sum |t_i(n_{j_2})| > 3\varepsilon/5.$$

Again if v_k is any complex number of modulus one for k other than

$N_1 + 1, \ldots, N_2$ then

$$\left|\sum v_i t_i(n_{j_2})\right| \geq \left|\sum_{N_1+1}^{N_2}\right| - \left|\sum_1^{N_1}\right| - \left|\sum_{N_2+1}^{\infty}\right| > \varepsilon/5.$$

This process can be repeated and, using induction, we can obtain a sequence $\{v_i\}$ such that $|v_i| = 1$ for all i, and so $\{v_i\} \in l_\infty$, and $|\sum v_i t_i(n_{j_k})| > \varepsilon/5$ for $k = 1, 2, \ldots$. This is a contradiction.

If $(B, \|\cdot\|)$ is a separable Banach space, there is a closed linear subspace G of $(l_1, \|\cdot\|_1)$ such that $(B, \|\cdot\|)$ and l_1/G are topologically isomorphic (Section 2.4, Theorem 1). Suppose that G has a complement, say H, in l_1. Let $\mathscr{F}$ be the family of all continuous, linear functionals on H that are restrictions to H of continuous, linear functionals on l_1. Then, by Lemma 1, a sequence $\{x_n\}$ of points of H converges to zero iff $\lim f(x_n) = 0$ for every $f \in \mathscr{F}$. It follows that there is a family, say $\mathscr{G}$, of continuous, linear functionals on $(B, \|\cdot\|)$ such that $\{y_n\} \subset B$ converges to zero iff $\lim g(y_n) = 0$ for every $g \in \mathscr{G}$. Hence, to prove that there is a closed, linear subspace of l_1 that does not have a complement it suffices to show that there is a separable Banach space $(B, \|\cdot\|)$ and a sequence $\{y_n\}$ of points of B such that $\lim g(y_n) = 0$ for every continuous, linear functional on $(B, \|\cdot\|)$, and yet $\{y_n\}$ is not convergent to zero. We shall show that the separable Banach space $(c_0, \|\cdot\|_\infty)$ has a sequence with these properties.

If $\{y_n\} \in l_1$ and $\{x_n\}$ is any element of c_0, then $|\sum x_n y_n| \leq \|\{x_n\}\|_\infty \|\{y_n\}\|_1$, and so every element of l_1 defines a continuous, linear functional on c_0. We have seen that $e_1, e_2, \ldots$ is a Schauder basis for c_0 (Exercises 1.3, problem 2). Let f be a continuous, linear functional on $(c_0, \|\cdot\|_\infty)$, let $M(f) = \sup\{|f(x)| \mid x \in c_0, \|x\|_\infty \leq 1\}$ and consider the sequence $\{f(e_n)\}$. For $\{x_n\} \in c_0, f(\{x_n\}) = \sum x_n f(e_n)$. Letting sgn $a = \bar{a}/|a|$ for $a \neq 0$, sgn$(0) = 0$, it is clear that $u_k = (\text{sgn } f(e_1), \text{sgn } f(e_2), \ldots, \text{sgn } f(e_k), 0, 0, \ldots)$ is in c_0 and has norm one. But for any k, $f(u_k) = \sum_1^k |f(e_n)| \leq M(f)$. Hence $\{f(e_n)\}$ is in l_1 and the space of continuous, linear functionals on c_0 is isomorphic to l_1.

Now, c_0 is certainly separable, and $\|e_n\| = 1$ for all n, yet $\lim f(e_n) = 0$ for every continuous, linear functional f on c_0.

Our work so far has clearly shown that the continuous, linear functionals on a normed space can be very useful in studying that space. In each of our examples there were always many obvious continuous, linear functionals on the given spaces. If, however, one considers a general normed space, then it is not clear, at this stage of our discussion, that there are any continuous, linear functionals on the space except the trivial one, i.e., the map that takes each point to zero. Let us ignore this difficulty for the moment.

Definition 1. Let $(E, \|\cdot\|)$ be a normed space. The vector space of all continuous, linear functionals on $(E, \|\cdot\|)$ will be called the dual space of $(E, \|\cdot\|)$ or, simply, the dual of E, and it will be denoted by E'.

If f is a continuous, linear functional on the normed space $(E, \|\cdot\|)$, then the number $\sup\{|f(x)| \,|\, x \text{ in } E, \|x\| \leq 1\}$ will be denoted by $\|f\|$. It is easy to see that the function that takes each f in E' to $\|f\|$ is a norm for E'. Whenever we speak of a norm on the dual of a normed space we shall always mean the norm defined in this way. This is sometimes called the dual space norm. Clearly, for any f in E', $|f(x)| \leq \|f\| \|x\|$ for all x in E.

Theorem 1. The dual of a normed space is always a Banach space.

Proof. Let $(E, \|\cdot\|)$ be a normed space and let $\{f_n\}$ be a Cauchy sequence in the normed space E'. Then given $\varepsilon > 0$ we can choose an integer N such that $\|f_n - f_m\| < \varepsilon$ whenever $m \geq N$ and $n \geq N$. For any x in E we have $|f_n(x) - f_m(x)| \leq \|f_n - f_m\| \|x\|$ and so $\{f_n(x)\}$ is a Cauchy sequence in the field K. Hence, for each x in E, we can let $f(x) = \lim f_n(x)$. Clearly, f is a linear functional on E and $|f(x) - f_m(x)| \leq \varepsilon \|x\|$ for all x, whenever $m \geq N$. Let $M = \sup\{\|f_n\| \,|\, n = 1, 2, \ldots\}$ (Exercises 1.2, problem 1). Then

$$|f(x)| \leq |f(x) - f_m(x)| + |f_m(x)| \leq (\varepsilon + M)\|x\|$$

for all x in E, and so $f \in E'$.

Finally,

$$\|f - f_m\| = \sup\{|f(x) - f_m(x)| \,|\, \|x\| \leq 1\} \leq \varepsilon$$

for $m \geq N$. This says that $\{f_n\}$ converges to f for the dual space norm.

It is sometimes possible to identify the dual of a given normed space (see problem 1 below). When we say that two normed spaces can be identified we mean that they are equivalent in the following sense:

Definition 2. Let $(E, \|\cdot\|)$ and $(F, \|\|\cdot\|\|)$ be two normed spaces over the same field. An isomorphism ϕ from E onto F is said to be an equivalence if $|\|\phi(x)\|| = \|x\|$ for all x in E. We shall say that two normed spaces are equivalent if there is an equivalence from one of these spaces onto the other.

EXERCISES 1

1. (a) Show that the dual of $(l_1, \|\cdot\|_1)$ is equivalent to $(l_\infty, \|\cdot\|_\infty)$.
    (b) Show that the dual of $(c_0, \|\cdot\|_\infty)$ is equivalent to $(l_1, \|\cdot\|_1)$.
2. (a) What can you say about any projection operator on $(l_\infty, \|\cdot\|_\infty)$ whose range is c_0?
    (b) What can you say about any supplement for c_0 in l_∞?
    (c) Let H be a subspace of l_∞ that is a supplement for c_0 in l_∞. What can you say about $\operatorname{cl} H \cap c_0$?
3. (a) Prove the "necessity" part of Lemma 1.
    (b) Show that the analog of Lemma 1 is true for any Banach space that is topologically isomorphic to $(l_1, \|\cdot\|_1)$.
    (c) Show that, if H is any linear subspace of $(l_1, \|\cdot\|_1)$, a sequence of points of H, $\{x_n\}$, converges to zero for the norm of H iff $\lim f(x_n) = 0$ for every continuous, linear functional on H.
4. (a) Referring to the paragraph just after Definition 1, show that the function defined there is a norm for E'.
    (b) If f is any element of E' prove that

    $$\|f\| = \sup\{|f(x)|\,\|x\|^{-1}\,|\,x \text{ in } E, x \neq 0\}$$
    $$= \sup\{|f(x)|\,|\,x \text{ in } E, \|x\| = 1\}.$$

*5. Let $(E, \|\cdot\|)$ and $(F, |\|\cdot\||)$ be two normed spaces and let $\mathscr{L}(E, F)$ be the space of all continuous, linear maps from E into F. If $T \in \mathscr{L}(E, F)$ let $\|T\|_l = \sup\{|\|T(x)\||\,|\,x \text{ in } E, \|x\| \leq 1\}$ (see Exercises 1.2, problem 3a).
    (a) Show that $\|\cdot\|_l$ is a norm for $\mathscr{L}(E, F)$.
    (b) If $(F, |\|\cdot\||)$ is a Banach space then show that $(\mathscr{L}(E, F), \|\cdot\|_l)$ is also a Banach space. Hint: Modify the proof of Theorem 1.
6. Let E be a vector space over K. A map ϕ from $E \times E$ into K is called an inner product on E if:

(i) $\phi(ax + by, z) = a\phi(x, z) + b(y, z)$ for all x, y, z in E and all scalars a, b;

(ii) $\phi(x, y) = \overline{\phi(y, x)}$ for all x, y in E;

(iii) $\phi(x, x) \geq 0$ for all x in E with equality iff $x = 0$. It is customary to denote $\phi(x, y)$ by $\langle x, y \rangle$.

(a) Let E be a vector space and let $\langle , \rangle$ be an inner product on E. Show that the function that takes x in E to the square root of $\langle x, x \rangle$ is a norm on E. This is the norm induced by the inner product.

(b) Let n be a positive integer and consider the vector space C^n. For $x = (x_1, \ldots, x_n)$ and $y = (y_1, \ldots, y_n)$ in this space let $\langle x, y \rangle = \sum_1^n x_k \bar{y}_k$. Show that $\langle , \rangle$ is an inner product on C^n.

(c) Show that the norm induced on C^n by the inner product defined in (b) is the Euclidian norm on C^n. Thus C^n is complete for this norm. If H is a vector space over K, $\langle , \rangle$ is an inner product on H and H is complete for the norm induced by this inner product, then we say that $(H, \langle , \rangle)$ is a Hilbert space. We shall not say very much about these spaces (but see Chapter 6). However, we must mention the following remarkable result: A Banch space is a Hilbert space iff each of its closed, linear subspaces has a complement [18].

2. The Hahn–Banach Theorem

Our long discussion of complementary subspaces began with the observation that such a pair of subspaces is associated with every compact operator (Section 2.2, Theorem 3). This led us to ask whether every closed, linear subspace of a Banach space has a complement. We answered that question in the last section. Let us go back now and recall that when we associated a pair of complementary subspaces with a compact operator one member of the pair was finite dimensional. So we may ask: Does every finite dimensional subspace of a Banach space have a complement? The answer is "yes," as we shall now show. Incidentally, the reader may wonder why we did not ask this more modest question right at the start (i.e., right after we proved Theorem 3 of Section 2.2). We could have, but the general question would have arisen sooner or later anyway.

Suppose that $(E, \|\cdot\|)$ is a normed space (there is no need to assume completeness, as we shall see) over the field K. Let G be a one-dimensional subspace of E. Then for any fixed, nonzero element x of G, $G = \{\lambda x \,|\, \lambda \text{ in } K\}$. Now assume that there is a continuous, linear functional ϕ on E such that $\phi(x) = 1$. If H denotes the null space of ϕ then H is certainly closed and $G \cap H = \{0\}$. For any y in E, $\phi(y)x$ is in G and $y - \phi(y)x$ is in H. Hence $y = \phi(y)x + [y - \phi(y)x]$, and we have shown that H is a complement for G in E.

It is easy to find a continuous, linear functional f on G such that $f(x) = 1$. Simply set $f(x) = 1$ and define f on all of G by linearity (Exercises 1.2, problem 3b). What we need is a continuous, linear functional ϕ on E such that $\phi(x) = 1$, i.e., a continuous, linear functional on E such that $\phi(y) = f(y)$ for all y in G. So we shall have shown that G has a complement in E once we have shown that every continuous, linear functional on G is the restriction to G of some continuous, linear functional on E. The next theorem proves this and more. It implies, in particular, that there are always plenty of continuous, linear functionals on any normed space. We stress that in the statement of the theorem G need not be finite dimensional.

Theorem 1 (Hahn–Banach Theorem).   Let $(E, \|\cdot\|)$ be a normed space over K and let G be a linear subspace of E. Given any continuous, linear functional f on G there is a continuous, linear functional f^* on E such that:

(i) $f(x) = f^*(x)$ for each x in G;
(ii) $\|f\| = \sup\{|f(x)| \,|\, x \in G, \quad \|x\| \le 1\} = \sup\{|f^*(y)| \,|\, y \in E, \|y\| \le 1\} = \|f^*\|.$

Proof.   Define a new norm on E, p, as follows: $p(y) = \|f\| \|y\|$ for each y in E. Observe that, for each x in G, $|f(x)| \le p(x)$. Now we distinguish two cases:

(a) Assume that K is R, the field of real numbers. Choose a point y in E but not in G and let V be the linear subspace of E generated by G and y. We shall show that there is a linear functional g on V that agrees with f on G and satisfies $|g(v)| \le p(v)$ for all v in V. Observe that, since $p(v) \ge 0$ for all v, we shall have proved this last inequality once we have shown that $g(v) \le p(v)$ for all v in V. Now $V = \{\lambda y + x \,|\, \lambda \text{ in } R, x \text{ in } G\}$ and g is to be linear, so g is completely determined once we define $g(y)$ since $g(\lambda y + x) = \lambda g(y) + f(x)$. The problem then, is to define $g(y)$ in such a way that the inequality $g(v) \le p(v)$ holds for every v in V.

If x_1, x_2 are in G, then

$$f(x_1) + f(x_2) = f(x_1 + x_2) \le p(x_1 + x_2)$$
$$= p[(x_1 - y) + (x_2 + y)] \le p(x_1 - y) + p(x_2 + y).$$

Hence

$$-p(x_1 - y) + f(x_1) \le p(x_2 + y) - f(x_2).$$

Since x_1 and x_2 are arbitrary elements of G we have, taking the sup and the inf over G, $\sup\{-p(x - y) + f(x)\} \le \inf\{p(x + y) - f(x)\}$. Thus we can choose a real number α that is between this supremum and infinum. Choose such an α and set $g(y) = \alpha$. We shall now prove that the functional defined in this way satisfies the required inequality.

If $\lambda > 0$,

$$g(\lambda y + x) = \lambda\alpha + f(x) = \lambda[\alpha + f(x/\lambda)]$$
$$\le \lambda[p(x/\lambda + y) - f(x/\lambda) + f(x/\lambda)]$$
$$= \lambda p(x/\lambda + y) = p(x + \lambda y).$$

If $\lambda = -\mu$, $\mu > 0$, then

$$g(-\mu y + x) = -\mu\alpha + f(x)$$
$$= \mu[-\alpha + f(x/\mu)] \le \mu[p(x/\mu - y) - f(x/\mu) + f(x/\mu)]$$
$$\le \mu p(x/\mu - y) = p(x - \mu y).$$

We have shown that we can extend f, in the required way, to subspaces of E that properly contain G. In order to extend f to all of E we must appeal to Zorn's lemma.

Let $\varepsilon = \{(V, g) \mid V$ is a linear subspace of E that contains G; g is a linear functional on V that agrees with f on G and satisfies $|g(v)| \le p(v)$ for all v in $V\}$. Partially order ε as follows: $(V_1, g_1) \le (V_2, g_2)$ iff $V_1 \subset V_2$ and g_2 agrees with g_1 on V_1. We now apply Zorn's lemma to prove that there is a maximal element (U, h) of ε. If $U \ne E$ then we can choose y in E but not in U and then extend h to a linear functional h^* on the linear subspace W generated by U and y. As we saw above this can be done in such a way that h and h^* agree on U and $|h^*(w)| \le p(w)$ for all w in W. But then (W, h^*) is in ε, $(U, h) \le (W, h^*)$ and $U \ne W$. This contradicts the maximality of (U, h).

We have proved the Hahn–Banach theorem for normed spaces over the field R. Before proving the theorem for normed spaces over C we must make some remarks about vector spaces over C.

If V is a vector space over C we can consider maps f from V into C such that $f(x + y) = f(x) + f(y)$ for all x, y in V and $f(\alpha x) = \alpha f(x)$ for all x in V and all real α. Such maps will be called *real linear functionals* on V. If f is a real linear functional on V whose range is contained in R then we shall say that f is a real-valued, real linear functional on V.

(b) Assume now that K is the field C. The given linear functional f on G can be written as follows: $f(x) = g(x) + ih(x)$, where g and h are real-valued, real linear functionals on G. We want to extend g and h to real-valued, real linear functionals g^* and h^* on E in such a way that the functional $g^* + ih^*$ extends f and satisfies our inequality. The first step is to observe that

$$g(ix) + ih(ix) = f(ix) = if(x) = i[g(x) + ih(x)].$$

Hence $h(x) = -g(ix)$ for all x in G. Since $|g(x)| \le |f(x)| \le p(x)$ on G there is, by (a), a real-valued, real linear functional g^* on E such that $|g^*(y)| \le p(y)$ for all y in E and $g^*(x) = g(x)$ for all x in G. Set $f^*(y) = g^*(y) - ig^*(iy)$ for all y in E. Then f^* is a complex-valued, complex linear functional on E and, for x in G,

$$f^*(x) = g^*(x) - ig^*(ix) = g(x) - ig(ix) = g(x) + ih(x) = f(x).$$

To establish our inequality we fix y in E and write $f^*(y) = re^{i\theta}$. Then

$$|f^*(y)| = r = e^{-i\theta}f^*(y) = f^*(e^{-i\theta}y)$$

and we can write:

$$|f^*(y)| = f^*(e^{-i\theta}y) = g^*(e^{-i\theta}y) = |g^*(e^{-i\theta}y)|$$
$$\le p(e^{-i\theta}y) = |e^{-i\theta}|p(y) = p(y).$$

Corollary 1.   Every finite dimensional subspace of a normed space has a complement.

Proof.   Let $(E, \|\cdot\|)$ be a normed space, let F be a finite dimensional subspace of E and let $x_1, \ldots, x_n$ be a basis for F. For each fixed j, $1 \le j \le n$, define $f_j(x_i) = 0$ for $i \ne j$, $f_j(x_j) = 1$, and extend f_j to all of F by linearity. Clearly each f_j is a continuous, linear functional on F (Exercises 1.2, problem 3b). Extend each f_j to a linear functional f_j^* on E that is continuous on $(E, \|\cdot\|)$, and let $N(f_j^*)$ be the null space of this extension. Clearly $G = \bigcap_{j=1}^{n} N(f_j^*)$ is a closed, linear subspace of E and $F \cap G = \{0\}$. To prove that G is a complement for F in E we need only show that $E = F + G$. But if $y \in E$ then $\sum_{j=1}^{n} f_j^*(y)x_j$ is in F and

$y - \sum_{j=1}^{n} f_j^*(y)x_j$ is in G. Since the sum of these elements is y, we are done.

Corollary 2. Let $(E, \|\cdot\|)$ be a normed space and let x_0 be any nonzero element of E. Then there is an element $f^* \in E'$ such that $\|f^*\| = 1$ and $f^*(x_0) = \|x_0\|$.

Proof. Let $G = \{\lambda x_0 \mid \lambda \text{ in } K\}$ and define a continuous, linear functional f on G as follows: $f(\lambda x_0) = \lambda\|x_0\|$ for all $x = \lambda x_0$ in G. Then $\|f\| = \sup\{|f(x)|\|x\|^{-1} \mid x \text{ is in } G, x \neq 0\} = \sup\{|\lambda|\|x_0\|\|\lambda x_0\|^{-1} \mid \lambda \text{ in } K, \lambda \neq 0\} = 1$. We may take f^* to be any "norm preserving" extension of f, i.e., any element of E' whose restriction to G is f and whose norm equals that of f.

Corollary 3. Let $(E, \|\cdot\|)$ be a normed space, let G be a closed, linear subspace of E, and let x_0 be a point of E that is not in G. Let $d = \inf\{\|x_0 - x\| \mid x \text{ in } G\}$. Then there is an element f^* in E' such that $\|f^*\| = 1, f^*(x_0) = d$, and $f^*(x) = 0$ for all x in G.

Proof. Set $V = \{x + \lambda x_0 \mid x \text{ in } G, \lambda \text{ in } K\}$, let $f(x_0) = d, f(x) = 0$ for all x in G, and define f at all other points of V by linearity. Once we have shown that f has norm one on V then we can take f^* to be any element of E' whose restriction to V is f and whose norm is one. Now $\|f\| = \sup\{|f(v)|\|v\|^{-1} \mid v \text{ in } V, v \neq 0\} = \sup\{|f(\lambda x_0 - x)|\|\lambda x_0 - x\|^{-1}\} = \sup\{|\lambda|d\|\lambda x_0 - x\|^{-1} \mid x \text{ in } G, \lambda \text{ in } K\} = \sup\{d\|x_0 - x\|^{-1} \mid x \text{ in } G\} = d/\inf\{\|x_0 - x\| \mid x \text{ in } G\} = d/d = 1$.

EXERCISES 2

1. Let $(E, \|\cdot\|)$ be a normed space over K. Let x be any nonzero element of E. Show that there is an element $f \in E'$ such that $f(x) \neq 0$. The null space of a linear functional on E is called a hyperplane in E (see Section 1.3, Lemma 1, and Exercises 1.3, problem 4). Show that any closed, linear subspace of $(E, \|\cdot\|)$ is the intersection of the closed hyperplanes that contain it.

2. Let $(E, \|\cdot\|)$ be a normed space and let $\{x_n\}$ be a sequence of points of E. We shall say that $\{x_n\}$ is a weak Cauchy sequence if $\{f(x_n)\}$ is

a Cauchy sequence for each f in E'. We shall say that $\{x_n\}$ is weakly convergent to the point x_0 of E if $\lim f(x_n) = f(x_0)$ for every f in E'. Finally, we shall say that $(E, \|\cdot\|)$ is weakly (sequentially) complete if every sequence of points of E that is a weak Cauchy sequence is weakly convergent to a point of E.

(a) If $\{x_n\}$ is weakly convergent to $x_0 \in E$ and also to $y_0 \in E$, show that $x_0 = y_0$.

(b) Show that a weakly complete normed space is a Banach space.

(c) Show that any closed, linear subspace of a weakly complete Banach space is itself weakly complete.

(d) If the Banach space $(E, \|\cdot\|)$ is weakly complete and if $(F, \|\|\cdot\|\|)$ is topologically isomorphic to $(E, \|\cdot\|)$, show that F is weakly complete.

(e) Show that $(l_1, \|\cdot\|_1)$ is weakly complete. Hint: Look over the proof of Lemma 1 (Section 1).

3.   The Banach–Steinhaus Theorem

A set $\mathcal{O}$ of continuous, linear functionals on a normed space $(E, \|\cdot\|)$ can be "bounded" in at least two different senses. First, since E' has a norm the set can be bounded for this norm; i.e., $\sup\{\|f\| \mid f \text{ in } \mathcal{O}\}$ can be finite. If this is the case then we shall say that $\mathcal{O}$ is a norm bounded set. It can also happen that $\sup\{|f(x)| \mid f \text{ in } \mathcal{O}\}$ is finite for each x in E. When this is true we shall say that $\mathcal{O}$ is pointwise bounded on E. A norm bounded set in E' is certainly pointwise bounded on E. The surprising fact is that, if $(E, \|\cdot\|)$ is complete, then the converse is true.

Theorem 1 (Banach–Steinhaus Theorem).   Let $(B, \|\cdot\|)$ be a Banach space. Then a subset of B' is norm bounded iff it is pointwise bounded on B.

Proof.   Let $\mathcal{O}$ be a subset of B' that is pointwise bounded on B. We may assume that $\mathcal{O}$ is countable, and so we may write $\mathcal{O} = \{f_n \mid n = 1, 2, \ldots\}$.

We shall now show that to prove that $\{f_n\}$ is norm bounded it suffices to prove that there is an x_0 in B, a $\delta > 0$, and a $k > 0$ such that

$|f_n(x)| < k$ for all n and all x in $\{x \in B \,|\, \|x - x_0\| \le \delta\}$. It is convenient to denote this set by $\mathcal{B}(\delta, x_0)$. Suppose that this condition is satisfied for an x_0 in B and numbers δ, k. Let x be a point of B with $\|x\| \le \delta$. Then

$$| f_n(x)| \le | f_n(x + x_0) - f_n(x_0)| \le | f_n(x + x_0)| + | f_n(x_0)| \le 2k$$

for every n because $\|(x + x_0) - x_0\| = \|x\| \le \delta$. Thus $|f_n(x)| \le 2k$ for every n and every x in the set $\mathcal{B}(\delta, 0)$. Now if y is any nonzero point of B, then $\delta y\|y\|^{-1}$ is in $\mathcal{B}(\delta, 0)$. So $|f_n(y)| \le 2k\|y\|\delta^{-1}$ for all n. Then clearly $\|f_n\| \le 2k\delta^{-1}$ for every n.

Now suppose that for any ball $\mathcal{S}$ (i.e., any set of the form $\mathcal{B}(\delta, x_0)$) and any $k > 0$ there is an integer n and a point x in $\mathcal{S}$ such that $|f_n(x)| \ge k$. Choose a ball $\mathcal{S}$, a point $x_1 \in \mathcal{S}$, and an integer $n(1)$ such that $|f_{n(1)}(x_1)| > 1$. Since $f_{n(1)}$ is continuous this inequality is satisfied at each point of a ball $\mathcal{S}_1$ contained in $\mathcal{S}$ and containing x_1 whose diameter is less than 2^{-1}. By our assumption there is a point $x_2 \in \mathcal{S}_1$ and an integer $n(2)$ such that $|f_{n(2)}(x_2)| > 2$. As before, this must hold at each point of some ball $\mathcal{S}_2$ contained in $\mathcal{S}_1$ and containing x_2 whose diameter is less than 2^{-2}. Now choose $x_3 \in \mathcal{S}_2$ and an integer $n(3)$ such that $|f_{n(3)}(x_3)| > 3$. Using the continuity of $f_{n(3)}$ we find a ball $\mathcal{S}_3$ contained in $\mathcal{S}_2$ and containing x_3 such that $|f_{n(3)}(x)| > 3$ for all $x \in \mathcal{S}_3$ and the diameter of $\mathcal{S}_3$ is less than 2^{-3}. Continue in this way. After $\mathcal{S}_1, \mathcal{S}_2, \ldots, \mathcal{S}_{k-1}$ have been chosen, choose $x_k \in \mathcal{S}_{k-1}$ and an integer $n(k)$ such that $|f_{n(k)}(x_k)| > k$. By the continuity of $f_{n(k)}$ this inequality must hold at each point of a ball $\mathcal{S}_k$ contained in $\mathcal{S}_{k-1}$ and containing x_k whose diameter is less than 2^{-k}.

Now we use the fact that $(B, \|\cdot\|)$ is a Banach space. Because of this, the set $\bigcap_{k=1}^{\infty} \mathcal{S}_k$ is not empty. But if y is a point in this intersection, then $|f_{n(k)}(y)| > k$ for $k = 1, 2, \ldots$, contradicting the fact that $\mathcal{O} = \{f_n\}$ is pointwise bounded on B.

This theorem has many important applications. We shall discuss some of these later on (Chapters 5 and 6). Right now let us show that the theorem need not be true for normed spaces that are not complete.

Let F be the linear subspace of l_1 consisting of all sequences that have only a finite number of nonzero terms. Clearly, F is dense in $(l_1, \|\cdot\|_1)$. For each fixed, positive integer n let $f_n(\{t_k\}) = t_n$ for each $\{t_k\}$ in F. Clearly, $f_n \in F'$ and $\|f_n\| \le 1$ for every n. Consider $\{nf_n\} \subset F'$. For any $\{t_k\}$ in F it is clear that $nf_n(\{t_k\}) = nt_n = 0$ for n sufficiently large. It follows that $\{nf_n\}$ is pointwise bounded on F. But $\|nf_n\| \ge |nf_n(e_n)| = n$, where e_n is the sequence having one in the nth position and zeros elsewhere. Thus $\{\|nf_n\|\}$ is not a bounded set.

EXERCISES 3

1. Referring to the first part of the proof of Theorem 1, why can we assume that the set $\mathcal{O}$ is countable?

2. Let $(B, \|\cdot\|)$ be a Banach space, let $(E, |\|\cdot\||)$ be a normed space, and let S be a subset of $\mathscr{L}(B, E)$ (Exercise 3.1, problem 5). If, for each x in B, the set $\{Tx \mid T \text{ in } S\}$ is bounded in $(E, |\|\cdot\||)$, show that there is a positive constant M such that $|\|Tx\|| \le M\|x\|$ for all x in B and each T in S.

3. Let $(B, \|\cdot\|)$ be a Banach space and let $\{f_n\}$ be a sequence in B'. Suppose that $\lim f_n(x)$ exists for each x in B and, for each x, define $g(x)$ to be this limit. Show that $g \in B'$.

4. Let $(E, \|\cdot\|)$ be a normed space and let S be a subset of E. Show that S is bounded in $(E, \|\cdot\|)$ iff $\sup\{|f(x)| \mid x \text{ in } S\}$ is finite for each f in E'.

4. The Completion of a Normed Space. Reflexive Banach Spaces

The dual of the normed space $(E, \|\cdot\|)$ is the Banach space E'. This space also has a dual. It is usually denoted by E'', is called the bidual of $(E, \|\cdot\|)$ and consists, of course, of all continuous, linear functionals on E'. For each fixed x in E define $\hat{x}(f)$ to be $f(x)$ for all f in E'. It is clear that $\hat{x}$ is a linear functional on E', and since $|\hat{x}(f)| = |f(x)| \le \|f\|\,\|x\|$ we see that $\hat{x}$ is in E''. Hence we can define a map ϕ from E into E'' by letting $\phi(x) = \hat{x}$ for each x in E. This map is linear and $\|\phi(x)\| = \|x\|$ for each x in E (Section 2, Corollary 2 to Theorem 1). So ϕ is an equivalence (Section 1, Definition 2) from $(E, \|\cdot\|)$ onto a linear subspace of E''. We shall call ϕ the canonical embedding of $(E, \|\cdot\|)$ into E'' and we shall often identify E with its image in E''. These observations have an immediate application.

Theorem 1.   Every normed space is equivalent to a dense, linear subspace of a Banach space.

Proof.   Let $(E, \|\cdot\|)$ be a normed space, let ϕ be the canonical embedding of $(E, \|\cdot\|)$ into E'', and let $\phi(E)$ be the image of E in E'' under ϕ. Clearly $(E, \|\cdot\|)$ and $\phi(E)$ are equivalent and $\phi(E)$ is a dense, linear subspace of its closure cl $\phi(E)$ in $(E'', \|\cdot\|)$. But $(E'', \|\cdot\|)$ is a Banach space (Section 1, Theorem 1). Hence cl $\phi(E)$ is also a Banach space.

Definition 1.   A Banach space is said to be a reflexive Banach space if the canonical embedding maps the space onto its bidual.

If B is a reflexive Banach space, then the canonical embedding is an equivalence from B onto its bidual. There are nonreflexive Banach spaces that are equivalent to their biduals [12]. Of course, in such a case, the equivalence is not the canonical map.

We have already seen that the dual of a separable Banach space need not be separable (Exercises 1, problem 1a). However:

Theorem 2.   Let $(E, \|\cdot\|)$ be a normed space and suppose that E' is a separable Banach space. Then $(E, \|\cdot\|)$ is separable.

Proof.   Let $S = \{f \text{ in } E' \mid \|f\| = 1\}$. We can choose a sequence f_1, $f_2, \ldots$ that is dense in S. For each n choose x_n in E with $\|x_n\| = 1$ and $|f_n(x_n)| > \frac{1}{2}$. Let G be the closure of the linear subspace of E generated by the set $\{x_n\}$. We shall show that $G = E$. Suppose that there is a point x_0 that is in E but not in G. By the third corollary to the Hahn–Banach theorem (Section 2) we can find an element g of E' such that $g(x_0) \neq 0$ but $g(x) = 0$ for all x in G. We can also assume that $\|g\| = 1$. Now

$$\tfrac{1}{2} < |f_n(x_n)| \leq |f_n(x_n) - g(x_n)| + |g(x_n)| \leq \|f_n - g\| \, \|x_n\| = \|f_n - g\|$$

for every n. But since $g \in S$ and $\{f_n\}$ is dense in this set we have reached a contradiction. So $G = E$ and, using this fact, we shall show that $(E, \|\cdot\|)$ is separable.

Suppose that E is defined over the field R. Let $\mathcal{G} = \{\sum_{j=1}^{n} \alpha_j x_j \mid n \text{ is finite}; \alpha_1, \alpha_2, \ldots, \alpha_n \text{ are in } R\}$ and let $\mathcal{H} = \{\sum_{j=1}^{n} q_j x_j \mid n \text{ is finite}; q_1, q_2, \ldots, q_n \text{ are rational numbers}\}$. It is clear that $\mathcal{G}$ is dense in G and that $\mathcal{H}$ is a countable set. Now given $\sum_{j=1}^{n} \alpha_j x_j$ in $\mathcal{G}$ and $\varepsilon > 0$ we may choose rational numbers $q_1, \ldots, q_n$ such that $|\alpha_j - q_j| < \varepsilon/n$ for $1 \le j \le n$. Then $\sum_{j=1}^n q_j x_j$ is in $\mathscr{H}$ and $\|\sum \alpha_j x_j - \sum q_j x_j\| < \varepsilon$. Hence $\mathscr{H}$ is a countable, dense subset of $(E, \|\cdot\|)$.

If E is defined over the field C the proof is similar to the one just given and we leave it to the reader.

EXERCISES 4

*1. Let $(E, \|\cdot\|)$ be a normed space. A Banach space that has a dense, linear subspace that is equivalent to $(E, \|\cdot\|)$ will be called a completion of $(E, \|\cdot\|)$.
     (a) Show that any two completions of $(E, \|\cdot\|)$ are equivalent.
     (b) If $(B, \|\cdot\|)$ is a completion of $(E, \|\cdot\|)$, show that B' and E' are equivalent.

2. To solve this problem one has to use some facts from real variable theory (see [9] or [21]).
    (a) Let $\mathscr{P}$ be the space of all polynomial functions on $[0, 1]$. For each f in $\mathscr{P}$ let $\|f\|_\infty = \sup\{|f(x)| \,|\, 0 \le x \le 1\}$. Identify the completion (up to equivalence, of course) of $(\mathscr{P}, \|\cdot\|_\infty)$.
    (b) For any fixed, real number $p \ge 1$ define a norm on $\mathscr{C}[0, 1]$ as follows: For each f in $\mathscr{C}[0, 1]$ let $\|f\|_p$ be the pth root of $\int_0^1 |f(x)|^p \, dx$. Identify the completion of $(\mathscr{C}[0, 1], \|\cdot\|_p)$.

3. (a) If $(B, \|\cdot\|)$ is a reflexive Banach space show that $(B', \|\cdot\|)$ is a reflexive Banach space.
    (b) If $(B, \|\cdot\|)$ is a separable, reflexive Banach space show that every dual space of $(B, \|\cdot\|)$ (i.e., B', B'', $(B'')'$, etc.) is a separable Banach space.
    (c) Show that $(c_0, \|\cdot\|_\infty)$ and $(l_1, \|\cdot\|_1)$ are not reflexive Banach spaces.

4. Let $(E, \|\cdot\|)$ be a normed space and let $f \in E'$ have norm one. For any $\varepsilon > 0$ show that there is an x_ε in E with $\|x_\varepsilon\| = 1$ and $f(x_\varepsilon) > 1 - \varepsilon$. Give an example to show that there need not be an x_0 in E such that $\|x_0\| = 1$ and $f(x_0) = 1$. Hint: Consider $(l_1, \|\cdot\|_1)$ and its dual $(l_\infty, \|\cdot\|_\infty)$.

CHAPTER 4

The Weak Topology

1. Topology from a Family of Seminorms

We come now to one of the most fascinating topics in the theory of normed spaces. This is the investigation of the weak topology and its relatives. In order to treat the subject properly we shall have to introduce new concepts and prove some peripheral, but interesting, results. However, most of this material will be used again, especially in the last chapter. An example of the kind of problem that we can solve using the ideas developed here is this: We know that l_∞ is the dual of l_1, and that l_1 is the dual of c_0. What about c_0; what space, if any, is it the dual of? More generally, given a Banach space B_1, is it always possible to find a Banach space B_2 such that B_1 and B'_2 are equivalent? This problem is solved in Chapter 5.

Definition 1. A real-valued function p on a vector space X is said to be a seminorm on X if:

(i) $p(x + y) \le p(x) + p(y)$ for all x, y in X.

(ii) $p(\alpha x) = |\alpha| p(x)$ for all x in X and all scalars α.

For example, if f is a linear functional on X then we can define a seminorm on X by setting $p_f(x) = |f(x)|$ for all x in X. In this way we obtain a whole family of seminorms on X.

Lemma 1. Let p be a seminorm on X. Then $p(0) = 0$ and $|p(x) - p(y)| \le p(x - y)$ for all x, y in X; in particular, $p(x) \ge 0$ for all x.

Proof. $p(0) = p(0x) = 0p(x) = 0$ by property (ii). By (i) we have $p(x - y) + p(y) \ge p(x)$ or $p(x - y) \ge p(x) - p(y)$. But

$$p(x - y) = |-1| p(y - x) \ge p(y) - p(x) = -\{p(x) - p(y)\}.$$

Hence we have our result.

Let X be a vector space and let $\{p_\gamma \mid \gamma$ in $\Gamma\}$ be a family of seminorms on X. There is a standard method of defining a topology on X by means of this family. We will give this method in three steps that will be referred to, collectively, as the construction process.

(a) Let $\mathscr{V}$ be the family of all subsets of X that are formed in the following way: Choose a finite subset $p_1, p_2, \ldots, p_n$ of $\{p_\gamma\}$, the same number of positive real numbers $\varepsilon_1, \varepsilon_2, \ldots, \varepsilon_n$, and let $V = V(p_1, \ldots, p_n; \varepsilon_1, \ldots, \varepsilon_n) = \{x$ in $X \mid p_j(x) < \varepsilon_j$ for $1 \le j \le n\}$.

(b) For any point x of X we shall say that a set U containing x is a neighborhood of x if there is some $V \in \mathscr{V}$ such that $x + V \subset U$.

(c) Let $t = t(\{p_\gamma\})$ be the family of all subsets of X that are neighborhoods of each of their points.

Lemma 2. The family of sets $t = t(\{p_\gamma\})$ is a topology on X and each p_γ is continuous on X for this topology. The topology is Hausdorff iff $\{p_\gamma\}$ satisfies the following: (Separation Condition) For each nonzero x in X there is some p_γ such that $p_\gamma(x) \ne 0$.

Proof. It is clear that X itself and the empty set are in t. We have to show that the union of any family of sets in t is a set in t, and that the intersection of any finite family of sets in t is a set in t. The first of these is clearly true. Let $\{O_j \mid j = 1, 2, \ldots, n\}$ be a finite family of sets in t. If the point x is in the intersection of these sets then for each j there is a set V_j

in $\mathscr{V}$ such that $x + V_j \subset O_j$. To prove that $\bigcap_{j=1}^n O_j$ is a neighborhood of x it is sufficient to prove that $\bigcap_{j=1}^n V_j$ is in $\mathscr{V}$. But this is obvious.

Choose any p_γ. It is clear, from Lemma 1, that p_γ is continuous on X iff it is continuous at zero. So given $\varepsilon > 0$ we have to show that for some neighborhood, say V, of zero, $p_\gamma(x) < \varepsilon$ for all x in V; i.e., we have to find some $V \in \mathscr{V}$ such that $p_\gamma(x) < \varepsilon$ for all x in V. But $\{x$ in $X \mid p_\gamma(x) < \varepsilon\}$ is in $\mathscr{V}$ by definition, and so p_γ is continuous at zero, hence on all of X, for the topology t.

Now let us prove the last statement. Assume that for $x \neq 0$ in X there is some p_γ such that $p_\gamma(x) \neq 0$. Then, if x, y are in X and $x \neq y$, there is some p_γ such that $p_\gamma(y - x) > 0$. So $p_\gamma(y) > p_\gamma(x)$. Let $d = p_\gamma(y) - p_\gamma(x)$ and let $V = \{z$ in $X \mid p_\gamma(z) < d/2\}$. We claim that $x + V$ and $y + V$ are neighborhoods of x and y, respectively, that are disjoint. To prove this let z belong to both of these sets. Then $z - x \in V$ and $z - y \in V$, and so $p_\gamma(z - x) < d/2$ and $p_\gamma(z - y) < d/2$. But then

$$p_\gamma(y - x) \le p_\gamma(y - z) + p_\gamma(z - x) < d,$$

which is a contradiction. So the separation condition does imply that t is Hausdorff. The converse is easy.

Remark. Let X be a vector space, let $\{p_\gamma\}$ be a family of seminorms on X, and let $t = t(\{p_\gamma\})$. The space X with the topology t will be denoted by $X[t]$. A continuous, linear functional on $X[t]$ is said to be a t-continuous, linear functional. Similarly, we shall speak of t-neighborhoods of zero in X, t-compact sets, t-convergent sequences, etc.

If $(E, \|\cdot\|)$ is a normed space and E' is its dual space, then the family of seminorms $\{p_f \mid f$ in $E'\}$ (recall that $p_f(x) = |f(x)|$ for all x in E) on E satisfies the separation condition (Section 3.2, Corollary 2 to the Hahn–Banach theorem). The Hausdorff topology $t(\{p_f\})$ will be denoted by $\sigma(E, E')$ and will be called the weak topology on E.

A typical $\sigma(E, E')$-neighborhood of zero contains a set of the form $\{x$ in $E \mid |f_j(x)| < \varepsilon_j$ for $1 \le j \le n\}$ (see (a) and (b) of the construction process). This last set contains a sufficiently small ball centered at zero; i.e., there is a $\delta > 0$ such that

$$V = \{x \text{ in } E \mid |f_j(x)| < \varepsilon_j \text{ for } 1 \le j \le n\} \supset \{x \text{ in } E \mid \|x\| < \delta\}.$$

Hence, by (b) and (c) of the construction process, every $\sigma(E, E')$-open subset of E is open for the norm topology. Notice that V contains $\bigcap_{j=1}^{n} \{x \text{ in } E \mid f_j(x) = 0\}$. This intersection is a linear subspace of finite codimension in E (Exercises 2.2, problem 6). Suppose that the norm and the weak topologies on E were to coincide. Then the unit ball of E, which is a norm neighborhood of zero, would have to contain a subspace of finite codimension in E. Now the only linear subspace of E that is contained in the unit ball is the zero subspace. This will have finite codimension in E iff E is finite dimensional. Hence we have proved:

(1)   *On an infinite dimensional normed space the weak topology is strictly weaker than the norm topology.*

By Lemma 2 each f in E' is $\sigma(E, E')$-continuous on E. Combining this with (1) we have:

(2)   *The space of all weakly continuous, linear functionals on a normed space E coincides with E'.*

If f is any element of E' and $N(f)$ is the null space of f, then $N(f)$ is a linear subspace of E that is closed for both the weak and the norm topologies. Let H be any norm closed, linear subspace of E. If $x \in E$, $x \notin H$, then there is, by Corollary 3 to the Hahn–Banach theorem, an element f of E' such that $f(x) \neq 0$ and $f(y) = 0$ for all y in H. Define $H^{\perp}$ to be $\{f \text{ in } E' \mid f \text{ vanishes on } H\}$. Then $H = \bigcap \{N(f) \mid f \text{ in } H^{\perp}\}$, and so H is closed for the weak topology; i.e., we have shown:

(3)   *A linear subspace of a normed space is closed for the weak topology iff it is closed for the norm topology.*

Now consider the dual space E' of the normed space E. This has a norm and a weak topology also (i.e., $\sigma(E', E'')$). There is, however, another useful topology on E'. For each fixed x in E define $p_x(f)$ to be $|f(x)|$ for all f in E'. The family of seminorms $\{p_x \mid x \text{ in } E\}$ defines a Hausdorff topology on E', which we shall call the weak* topology on E'. This topology is written $\sigma(E', E)$.

Since $E \subset E''$ (Section 3.4) the construction process shows that $\sigma(E', E)$ is weaker than $\sigma(E', E'')$. We can say more. But first we shall prove:

(4)   *The space of all weak* continuous, linear functionals on E' coincides with E.*

To prove this let ϕ be a $\sigma(E', E)$-continuous, linear functional on

E'. Then $\{f \in E' \mid |\phi(f)| < 1\}$ contains a $\sigma(E', E)$-neighborhood of zero. Hence there is a finite set $x_1, x_2, \ldots, x_n$ in E and positive numbers $\varepsilon_1, \varepsilon_2, \ldots, \varepsilon_n$ such that $|\phi(f)| < 1$ whenever f is in $\{g \in E' \mid |g(x_j)| < \varepsilon_j$ for $1 \le j \le n\}$. In particular, if $N(\hat{x}_j)$ is the null space in E' of the functional $\hat{x}_j$ (see Section 3.4 for the notation $\hat{x}_j$), $|\phi(f)| < 1$ whenever $f \in \bigcap_{j=1}^n N(\hat{x}_j)$. But if f is in this intersection, then so is λf for all scalars λ. Thus $|\phi(\lambda f)| < 1$ for all λ and this implies $\phi(f) = 0$, i.e., $N(\phi) \supset \bigcap_{j=1}^n N(\hat{x}_j)$. It follows that ϕ is a linear combination of $x_1, \ldots, x_n$ (Exercises 2.2, problem 6) and so $\phi \in E$.

Since the space of all $\sigma(E', E'')$-continuous, linear functionals on E' coincides with E'' (by (2)) we have:

(5) *The topology $\sigma(E', E)$ is weaker than $\sigma(E', E'')$ and when $E \ne E''$ these topologies are distinct.*

A linear subspace of E' that is norm closed need not be weak* closed, but we can say something about its weak* closure. We need some notation first. For any linear subspace M of E' let $M_\perp = \{x$ in $E \mid f(x) = 0$ for every f in $M\}$ and let $(M_\perp)^\perp$ be $\{f$ in $E' \mid f(x) = 0$ for all x in $M_\perp\}$. It is clear that both $M_\perp$ and $(M_\perp)^\perp$ are linear subspaces. Also, since $M_\perp = \bigcap \{N(f) \mid f$ in $M\}$ and $(M_\perp)^\perp = \bigcap \{N(\hat{x}) \mid x$ in $M_\perp\}$, $M_\perp$ is closed in E and $(M_\perp)^\perp$ is weak* closed in E'.

Lemma 3. The weak* closure of any linear subspace of M of E' coincides with $(M_\perp)^\perp$.

Proof. Since $M \subset (M_\perp)^\perp$ all we have to do is show that if f_0 is not in the weak* closure of M then it is not in $(M_\perp)^\perp$. Assume that f_0 in E' is not in the weak* closure of M. We shall find a point x_0 of $M_\perp$ such that $f_0(x_0) \ne 0$. Our assumption implies that there is a weak* neighborhood of f_0 that does not meet M; i.e., there is a finite set $x_1, \ldots, x_n$ in E and $\varepsilon > 0$ such that no point $g \in M$ can satisfy $|g(x_i) - f_0(x_i)| < \varepsilon$ for $1 \le i \le n$. Define a map ϕ from E' into K^n by letting $\phi(g) = (g(x_1), \ldots, g(x_n))$ for every $g \in E'$. Clearly $\phi(M)$ is a linear subspace of K^n and $\phi(f_0)$ is not in this subspace. Thus there is a linear functional on K^n that is zero on $\phi(M)$ and is not zero at $\phi(f_0)$ (by the Hahn–Banach theorem, Corollary 3); i.e., there are constants $c_1, c_2, \ldots, c_n$ such that $\sum c_i g(x_i) = 0$ for all $g \in M$ and $\sum c_i f_0(x_i) \ne 0$ (Exercises 1.3, problem 5). Let $x_0 = \sum c_i x_i$ and note that our last statement becomes: $g(x_0) = 0$ for all g in M (hence $x_0 \in M_\perp$) and $f_0(x_0) \ne 0$ (hence $f_0 \notin (M_\perp)^\perp$).

Suppose that ϕ is in E'' but is not in E. Then $N(\phi) = \{f$ in $E' | \phi(f) = 0\}$ is a norm closed, linear subspace of E' whose $\sigma(E', E)$-closure is $(N(\phi)_\perp)^\perp$. Now $N(\phi)_\perp = \{x$ in $E | f(x) = 0$ for every f in $N(\phi)\}$. If $x \in E$ is in this set then $x = \lambda\phi$ (Exercises 1.3, problem 4). But $\phi \notin E$ and so this is possible only for $\lambda = 0$; i.e., $N(\phi)_\perp$ is the linear subspace of E whose only element is the zero vector. Clearly then $(N(\phi)_\perp)^\perp$ is all of E' and so $N(\phi)$ is $\sigma(E', E)$-dense in E'. We have proved:

(6)   If $E \neq E''$, then there are linear subspaces of E' that are both norm closed and weak* dense in E'.

The next theorem exhibits another important difference between the weak and weak* topologies. This result has many applications and we shall discuss some of these later on.

Theorem 1 (Alaoglu's Theorem).   The unit ball, in the dual of any normed space, is compact for the weak* topology.

Proof. Let $(E, \|\cdot\|)$ be a normed space over K and, for each x in E, let $K(x)$ be K. An element θ of $\prod \{K(x) | x$ in $E\}$ is a function from E, the index set, into K. Hence, it makes sense to speak of $\theta(x)$. Every element of E' can be regarded as a point in the product space; i.e., there is a map from E' into $\prod \{K(x) | x$ in $E\}$. Identify E' with its image in the product space under this map. Notice that, since E is the index set for the product space, the product topology restricted to E' is just $\sigma(E', E)$.

Let $\mathscr{B}'$ be the unit ball of E' and, for each x in E, let $D(x) = \{z$ in $K \mid |z| \leq \|x\|\}$. If $g \in \mathscr{B}'$ then, for any x, $|g(x)| \leq \|g\| \, \|x\| \leq \|x\|$, and so $\mathscr{B}'$ is contained in the compact set $\prod \{D(x) | x$ in $E\}$. If we can show that $\mathscr{B}'$ is closed in this product, we shall be done. Let θ be a point in $\prod \{D(x) | x$ in $E\}$ that is in the closure of $\mathscr{B}'$. The first thing we shall do is show that $\theta \in E'$. Given $\varepsilon > 0$ and x, y in E we can find $g \in \mathscr{B}'$ such that

$$|g(x) - \theta(x)| < \varepsilon, \qquad |g(y) - \theta(y)| < \varepsilon,$$

and

$$|g(x + y) - \theta(x + y)| < \varepsilon$$

because θ is in the closure of $\mathscr{B}'$ for the product space topology. Now g is linear and so we get $|\theta(x + y) - \theta(x) - \theta(y)| < 3\varepsilon$ and, since ε was arbitrary, $\theta(x + y) = \theta(x) + \theta(y)$ for all x, y in E. In a similar way one can show that $\theta(\alpha x) = \alpha\theta(x)$ for all α in K and all x in E. So θ is linear. Recall now that $\theta \in \prod \{D(x) \,|\, x \text{ in } E\}$. This means $|\theta(x)| \leq \|x\|$ for all x in E and so the linear map θ must be continuous on E. In fact, $\theta \in \mathscr{B}'$.

For a reflexive Banach space (Section 3.4, Definition 1) B the weak and weak* topologies on B' coincide; i.e., $\sigma(B', B) = \sigma(B', B'')$ because $B = B''$. More is true in this case. Since $B = B''$ we must also have $B' = B'''$. Hence $\sigma(B, B')$ and $\sigma(B'', B''') = \sigma(B'', B')$ coincide on B. This says that, when B is a reflexive Banach space, the weak topology on B is actually a weak* topology. Applying Alaoglu's theorem we have:

Corollary 1. The unit ball of a reflexive Banach space is compact for the weak topology.

We shall investigate the converse of Corollary 1 later on.

EXERCISES 1

*1. Let X be a vector space over K and let $\{p_\gamma\}$ be a family of semi-norms on X. We refer to step (a) of the construction process.
    (a) Show that, if $U \in \mathscr{V}$, there is a set $V \in \mathscr{V}$ such that $V + V \subset U$.
    (b) If $U \in \mathscr{V}$ show that there is a set $V \in \mathscr{V}$ and a neighborhood N of zero in K such that $\alpha V \subset U$ for all α in N.
    (c) Let $t = t(\{p_\gamma\})$. Show that a sequence $\{x_n\} \subset X$ is t-convergent to $x \in X$ iff $\lim p_\gamma(x - x_n) = 0$ for each γ.

*2. Let $(E, \|\cdot\|)$ be a normed space. We refer to Exercises 3.2, problem 2.
    (a) Show that $\{x_n\} \subset E$ is a weak Cauchy sequence iff for every $\sigma(E, E')$-neighborhood V of zero in E there is an integer N such that $x_n - x_m \in V$ whenever both $m, n \geq N$.
    (b) Show that $\{x_n\} \subset E$ is weakly convergent to $x \in E$ iff $\lim f(x - x_n) = 0$ for every f in E'.

*3. Let $(E, \|\cdot\|)$ be a normed space and let S be a subset of E. We shall say that S is a total subset of E if $f \in E'$ and $f(x) = 0$ for all $x \in S$ implies $f = 0$.

(a)   Show that S is a total subset of E iff the linear subspace generated by S is dense in E.

(b)   Show that $(E, \|\cdot\|)$ is separable iff it contains a countable, total subset.

*4.   Let $(E, \|\cdot\|)$ and $(F, \|\|\cdot\|\|)$ be two normed spaces and suppose that u is a linear map from E into F. Define a map u^* from F' into E' as follows: For each $\phi \in F'$, $u^*(\phi)$ is the element of E' defined by $u^*(u)(x) = \phi[u(x)]$ for all x in E. This is called the adjoint of u.

(a)   Show that u^* is linear. If u is continuous prove that u^* is continuous when E' and F' have their norm topologies. If u is continuous show that $\|u\| = \|u^*\|$ and also that u^* is continuous when E' and F' have their weak* topologies. Hint: For each fixed $x \in E$ we have

$$| \|u(x)\| | = \sup\{ |\phi[u(x)]| \mid \phi \in F', \|\phi\| \le 1\}.$$

But $\phi[u(x)] = u^*(\phi)(x)$ by definition. Thus $\|u(x)\| \le \sup\{ |u^*(\phi)(x)| \mid \phi \in F', \quad \|\phi\| \le 1\}$. It follows that $| \|u(x)\| | \le \|u^*\| \|x\|$, and so $\|u\| \le \|u^*\|$. The reverse inequality is easy to prove.

(b)   If u is an equivalence from E onto F (Section 3.1, Definition 2) show that u^* is an equivalence.

*5.   Let $(E, \|\cdot\|)$ be a separable normed space. Show that any sequence in E' that is bounded for the norm of E' has a subsequence that is $\sigma(E', E)$-convergent. Hint: Let $\{f_n\}$ be a bounded sequence in the Banach space E' and let $\{x_j\}$ be a countable, dense subset of $(E, \|\cdot\|)$. For each fixed j the set $\{f_n(x_j) \mid n = 1, 2, \ldots\}$ is bounded in K and hence has a convergent subsequence. For an interesting and useful application of this result see [27, Theorem 2b, p. 103].

2.   Sets Which Define Seminorms

Seminorms arise in another, more geometric, way. Let X be a vector space over the field K and let A be a subset of X that contains the zero vector. Suppose that $x \in X$ and that x is in $\alpha A = \{\alpha y \mid y \in A\}$ for some positive scalar α. In this case we define $p_A(x)$ to be $\inf\{\alpha > 0 \mid x$ is in $\alpha A\}$. If x is never in a multiple of A then we set $p_A(x) = +\infty$. The function p_A from x into $R \cup \{+\infty\}$ is called the gauge function of A. Notice that if $\lambda \geq 0$, $p_A(\lambda x) = \lambda p_A(x)$ for any x.

Definition 1. A subset A of a vector space X is said to be a *convex set* if for any two points x, y of A and any real number α, $0 \leq \alpha \leq 1$, the point $\alpha x + (1 - \alpha)y$ is in A.

Lemma 1. Let A be a convex set that contains the zero vector and let p_A be the gauge function of A. Then $p_A(x + y) \leq p_A(x) + p_A(y)$ for all x, y.

Proof. For any positive numbers λ and μ we have $(\lambda + \mu)A \subset \lambda A + \mu A$. Since A is convex,

$$\left(\frac{\lambda}{\lambda + \mu}\right)A + \left(\frac{\mu}{\lambda + \mu}\right)A \subset A.$$

Hence, for a convex set A, $(\lambda + \mu)A = \lambda A + \mu A$.

Our inequality is certainly satisfied if $p_A(x)$ or $p_A(y)$ is $+\infty$. Assume that $p_A(x) = s$ and $p_A(y) = t$, where both are finite, and let $\varepsilon > 0$ be given. We can choose λ, μ such that

$$s \leq \lambda < s + \varepsilon, \qquad x \text{ is in } \lambda A$$

and

$$t \leq \mu < t + \varepsilon, \qquad y \text{ is in } \mu A.$$

It follows from this that $x + y$ is in $\lambda A + \mu A = (\lambda + \mu)A$, and so $p_A(x + y) \leq s + t + 2\varepsilon$. The inequality follows from this.

Definition 2. Let A be a subset of the vector space X.

(a) We shall say that A is a *balanced set* if $\alpha A \subset A$ for all scalars α such that $|\alpha| \leq 1$.

(b) We shall say that A is an *absorbing set* if for any x in X there is some scalar $\alpha > 0$ such that αA contains x.

Observe that a balanced set always contains the zero vector and that the gauge function of an absorbing set is a real-valued function.

Lemma 2. The gauge function of a balanced set A satisfies: $p_A(\lambda x) = |\lambda| p_A(x)$ for all scalars λ and all vectors x.

Proof. Since the equation is true for all $\lambda > 0$ we need only prove it for scalars λ such that $|\lambda| = 1$. But if $|\lambda| = 1$ then, since A is balanced, a vector $x \in A$ iff it belongs to λA.

The following theorem summarizes our results so far.

Theorem 1. Let X be a vector space over K. The gauge function of an absorbing, balanced, convex subset of X is a seminorm on X. Conversely, if p is a seminorm on X, then the set $A = \{x \mid p(x) < 1\}$ is an absorbing, balanced, convex set such that $p_A = p$.

EXERCISES 2

1. Let X be a vector space over K.
    (a) Prove that the intersection of any family of convex (respectively, balanced) subsets of X is a convex (respectively, balanced) subset of X.
    *(b) Given a subset S of X define the convex hull of S to be the intersection of the family of all convex subsets of X that contain S. Show that the convex hull of S is $\{\sum_{j=1}^{n} \alpha_j x_j \mid n$ is a positive integer; $x_1, \ldots, x_n$ are in S; $\alpha_1, \ldots, \alpha_n$ are nonnegative numbers with $\sum_{j=1}^{n} \alpha_j = 1\}$.
    *(c) Define the balanced, convex hull of S to be the intersection of the family of all balanced, convex subsets of X that contain S. Show that this is equal to $\{\sum_{j=1}^{n} \alpha_j x_j \mid n$ is a positive integer; $x_1, \ldots, x_n$ are in S; $\alpha_1, \ldots, \alpha_n$ are elements of K with $\sum_{j=1}^{n} |\alpha_j| \leq 1\}$.
    (d) Define the balanced hull of S in the natural way (see (b) and (c)). Show that the balanced, convex hull of S is equal to the convex hull of the balanced hull of S, but that it need not equal the balanced hull of the convex hull of S.

2. Let $(E, \|\cdot\|)$ be a normed space over K, and let $\mathscr{B}$ be the unit ball of this space.
    (a) Show that a balanced, closed, convex subset C of $(E, \|\cdot\|)$ is the unit ball corresponding to some norm on E that is weaker than $\|\cdot\|$ if, for some $\alpha > 0$, $\alpha\mathscr{B} \subset C$.
    (b) If, in addition to the conditions on C stated in (a), we

assume that the set is bounded, then show that the gauge function of C is a norm on E that is equivalent to $\|\cdot\|$.

(c) Let $(B, \|\cdot\|)$ be a Banach space and let C be an absorbing balanced, bounded, closed, convex subset of B. Show that the gauge function of C is a norm on B that is equivalent to $\|\cdot\|$.

3. Locally Convex Spaces. Kolmogorov's Theorem

Here we shall examine the topology defined by a family of seminorms in some detail. Throughout this section X will be a vector space over K, $\{p_\gamma \,|\, \gamma \text{ in } \Gamma\}$ a family of seminorms on X, and t the topology $t(\{p_\gamma\})$. We invite the reader to look over the construction process given just after Definition 1 in Section 1. In particular, recall that any t-neighborhood of, say $x_0 \in X$, contains a set of the form $x_0 + V$, where $V \in \mathscr{V}$.

Lemma 1. We consider the space $X[t]$ and we give $X \times X$, $K \times X$ their respective product space topologies. Then:

(i) The map from $X \times X$ into X that takes each pair (x, y) onto $x + y$ is continuous.

(ii) The map from $K \times X$ into X that takes each pair (α, x) onto αx is continuous.

Proof. If x_0, y_0 in X and a t-neighborhood W of $x_0 + y_0$ are given, we can find $U \in \mathscr{V}$ such that $x_0 + y_0 + U \subset W$. By problem 1a of Exercises 1, there is a $V \in \mathscr{V}$ such that $V + V \subset U$. Now $(x_0 + V) \times (y_0 + V)$ is a neighborhood of (x_0, y_0) in $X \times X$ and our map takes the neighborhood to $x_0 + V + y_0 + V \subset x_0 + y_0 + U \subset W$. This proves (i).

Now let $\alpha_0 \in K$, $x_0 \in X$, and a neighborhood W of $\alpha_0 x_0$ be given. We want to find a neighborhood N of α_0 in K and a neighborhood $U_2(x_0)$ of x_0 in X such that $(\alpha x - \alpha_0 x_0) \in W$ whenever $(\alpha, x) \in N \times U_2(x_0)$. There is a set $V \in \mathscr{V}$ such that $\alpha_0 x_0 + V \subset W$, and we can find another set $U \in \mathscr{V}$ such that $U + U + U \subset V$ (Exercises 1, problem 1a). The set U is balanced and absorbing (Section 2, Theorem 1) and so

(1) if $x - x_0 \in U$, then $\alpha(x - x_0) \in U$ for all α with $|\alpha| \le 1$;

(2) there is a number $\varepsilon > 0$ such that $\alpha x_0 \in U$ for all α with $|\alpha| \leq \varepsilon$.

Suppose (we shall prove this in a moment) that we can find a set $U_1 \in \mathscr{V}$ such that

(3) $\quad \alpha_0 U_1 \subset U,$

and let U_2 be any element of $\mathscr{V}$ that is contained in $U_1 \cap U$. Set $N = \{\alpha \in K \,|\, |\alpha - \alpha_0| \leq \min(1, \varepsilon)\}$ and set $U_2(x_0) = x_0 + U_2$. If $(\alpha, x) \in N \times U_2(x_0)$, then $(\alpha - \alpha_0)x_0 \in U$ by (2), $\alpha_0(x - x_0) \in U$ by (3) and the fact that $(x - x_0) \in U_2 \subset U_1$, and $(\alpha - \alpha_0)(x - x_0) \in U$ by (1) and the fact that $(x - x_0) \in U_2 \subset U$. But

$$\alpha x - \alpha_0 x_0 = (\alpha - \alpha_0)x_0 + \alpha(x - x_0) + (\alpha - \alpha_0)(x - x_0)$$

and so $\alpha x - \alpha_0 x_0$ is in $U + U + U \subset V$.

We shall now complete the proof by showing that for any $V \in \mathscr{V}$ and any $\alpha_0 \in K$ there is a set $U \in \mathscr{V}$ such that $\alpha_0 U \subset V$. We know that there is a set $U \in \mathscr{V}$ with $U + U \subset V$. Hence, by induction, we can find, for any integer n, a set $U_n \in \mathscr{V}$ such that $2^n U_n \subset V$. Choose n so that $|\alpha_0| \leq 2^n$. Then, since U_n is a balanced set, $\alpha_0 2^{-n} U \subset U$. Thus $\alpha_0 U \subset 2^n U \subset V$.

Definition 1. Let X be a vector space over K and let s be any topology on X. We shall say that s is compatible with the vector space structure of X and that $X[s]$ is a topological vector space, if X with the topology s satisfies conditions (i) and (ii) of Lemma 1.

A normed space with its norm topology is a topological vector space; this is, since a norm is a seminorm, a special case of Lemma 1. Observe that if $X[s]$ is a topological vector space and x_0 is any fixed point of X, the map that takes each x in X to $x + x_0$ is a homeomorphism. Hence the s-neighborhoods of any point are just translates of the s-neighborhoods of zero and, just as for normed spaces, we can compare compatible topologies on a vector space by comparing the neighborhoods of zero in these topologies.

Now the only topological vector spaces that we have seen are those whose topologies could be defined by means of a family of seminorms. We may ask whether the topology of every topological vector space arises in this way. More precisely, given any topological vector space $X[s]$, does there exist a family of seminorms $\{p_\gamma\}$ on X such that $s = t(\{p_\gamma\})$? This is what we will mean if we say that the topology on a certain $X[s]$ can be defined by a family of seminorms. The answer to our question is, in general, "no" [16, p. 156], but we can give a nice characterization of those spaces for which the answer is "yes."

Definition 2.   A topological vector space $X[s]$ is said to be a locally convex space if every s-neighborhood of zero in X contains a convex s-neighborhood of zero.

It is easy to see that if t is defined by a family of seminorms, then $X[t]$ is a locally convex space.

Lemma 2.   Let $X[s]$ be a locally convex space. Then every s-neighborhood of zero in X contains an s-neighborhood of zero that is an absorbing, balanced, convex, s-open set.

Proof.   Let U be any s-neighborhood of zero. Clearly, U contains an open s-neighborhood of zero (the interior of U, for example), so let us assume that U is open. If x_0 is in X then clearly $0x_0 \in U$ and hence there is a neighborhood N of 0 in K such that $\alpha x_0 \in U$ for all α in N; i.e., there is a positive number β' such that $\alpha x_0 \in U$ if $|\alpha| < \beta'$. From this we see immediately that U is absorbing since $x_0 \in (1/\alpha)U$ whenever α is a positive number that is less than β'.

Now the zero vector is in U and so there is an s-neighborhood V of zero in X and a neighborhood N of zero in K such that $N \times V \subset U$. Clearly N can be taken to be $\{\alpha \text{ in } K \,|\, |\alpha| < \beta''\}$ for some positive number β''. Choose, and fix, a positive number β such that $\beta < \beta''$. Then $\alpha V \subset U$ if $|\alpha| \leq \beta$. Then choose an s-neighborhood V_0 of zero such that $(1/\beta)V_0 \subset V$. If α is in K and $|\alpha| \leq 1$, then $\alpha V_0 = \alpha\beta(1/\beta)V_0 \subset (\alpha\beta)V \subset U$ because $|\alpha\beta| \leq \beta$. Thus the balanced s-neighborhood $\bigcup \{\alpha V_0 \,|\, |\alpha| \leq 1\}$ is in U.

We have seen that U is absorbing and that U contains a balanced s-neighborhood of zero; call it U_1. Since $X[s]$ is a locally convex space we can find a convex s-neighborhood U_2 of zero that is contained in U_1. If V is the balanced, convex hull of U_2 (Exercises 2, problem 1c), then V is an absorbing, balanced, convex s-neighborhood of zero that is contained in U. But the interior of V also has these properties and it is clearly an open set. This proves the lemma.

Theorem 1.   The family of all topological vector spaces whose topologies can be defined by a family of seminorms coincides with the family of all locally convex spaces.

Proof. We need only prove the sufficiency of this condition. Let $X[s]$ be a locally convex space and let $\mathscr{V}$ be the family of all absorbing, balanced, convex, open neighborhoods of zero in X. For each V in $\mathscr{V}$ let P_V be the gauge function of V, and let $t = t(\{P_V \,|\, V \text{ in } \mathscr{V}\})$. Then $X[t]$ is a locally convex space. We shall show now that $s = t$.

Let $\mathscr{V}(t)$ denote the family of sets defined in part (a) of the construction process. If W is any t-neighborhood of zero in X then W contains a set $V \in \mathscr{V}(t)$. There is a finite set $V_1, \ldots, V_n$ in $\mathscr{V}$ and numbers $\varepsilon_1, \ldots, \varepsilon_n$ such that $V = \{x \text{ in } X \,|\, P_{v_j}(x) < \varepsilon_j \text{ for } 1 \le j \le n\}$. Now

$$\{x \,|\, P_{v_j}(x) < \varepsilon_j\} = \varepsilon_j \{x \,|\, P_{v_j}(x) < 1\} = \varepsilon_j V_j$$

so $V = \bigcap \varepsilon_j V_j$ is in $\mathscr{V}$. This proves that t is weaker than s. The fact that s is weaker than t is proved similarly.

The normed spaces are a subfamily of the family of all Hausdorff, locally convex spaces. This subfamily was neatly characterized by Kolmogorov, and we shall present his characterization now.

Definition 3.   Let $X[t]$ be a locally convex space. A subset B of X is said to be a t-bounded set if, for every t-neighborhood U of zero, there is a scalar λ such that $B \subset \lambda U$.

Theorem 2.   Let $X[t]$ be a locally convex space. There is a norm on X such that the norm topology is equivalent to t iff there is a t-bounded t-neighborhood of zero in X.

Proof.   Suppose that there is a norm $\|\cdot\|$ on X such that the norm topology is equivalent to t. Then, if $\mathscr{B}$ is the unit ball of $(X, \|\cdot\|)$, $\mathscr{B}$ is a t-neighborhood of zero. But since t is equivalent to the norm topology any t-neighborhood V of zero must contain some multiple of $\mathscr{B}$; i.e., there is a $\mu > 0$ such that $V \supset \mu\mathscr{B}$. Clearly then, $(1/\mu)V \supset \mathscr{B}$ and $\mathscr{B}$ is a t-bounded set.

Now suppose that $X[t]$ contains a t-bounded t-neighborhood of zero. We may assume that W is balanced, t-closed, and convex. The gauge function q of W is a seminorm on X. Suppose that, for some $x \in X$, $q(x) = 0$. Then for each positive integer n there is an $x_n \in W$ such that $x_n = nx$. Let p be any t-continuous seminorm on X, let $U = \{x \text{ in } X \,|\, p(x) < 1\}$, and choose λ in K such that $W \subset \lambda U$. Then $p(x) < \lambda$ for all x in W. Combining these observations we have

$p(x) < \lambda/n$ for every n. It follows that $p(x) = 0$ and, since t is a Hausdorff, locally convex topology, $x = 0$. Thus q is a norm on X. We must now show that the topology induced on X by q is equivalent to t. This is easy. Since W is a t-neighborhood of zero t is stronger than the norm topology, but since W is bounded t is weaker than this topology.

EXERCISES 3

*1. Let $X[t]$ be a locally convex space and let f be a linear functional on X.
  (a) If x_0 is a fixed point of X show that the map that takes each point x of X to $x + x_0$ is a homeomorphism in $X[t]$.
  (b) Show that f is t-continuous on X iff it is t-continuous at zero. Show that f is t-continuous at zero iff there is a t-neighborhood U of zero such that $\sup\{|f(x)| \,|\, x \text{ in } U\}$ is finite.
  (c) Assume that $f \neq 0$ and show that the following conditions on f are equivalent:
    (i) f is t-continuous on X.
    (ii) The null space of f is a proper, t-closed, linear subspace of X.
    (iii) The null space of f is not t-dense in X. (Hint: See the proof of Lemma 1 in Section 1.3.)
  (d) Suppose that $(B, \|\cdot\|)$ is a nonreflexive Banach space. Using (c) show that there are linear subspaces of B' that are norm closed but $\sigma(B', B)$-dense in B'.
2. Let $X[t]$ be a locally convex space. Show that every t-neighborhood of zero in X contains an absorbing, balanced, convex, t-closed t-neighborhood of zero.
3. Let $X[t]$ be a locally convex space, let B be a subset of X, and let $t = t(\{p_\gamma\})$, where $\{p_\gamma \,|\, \gamma \text{ in } \Gamma\}$ is a family of seminorms on X.
  *(a) Show that B is t-bounded iff $\sup\{p_\gamma(x) \,|\, x \text{ in } B\}$ is finite for every γ.
  (b) If B is t-bounded show that the t-closure of B is t-bounded.
  *(c) If ϕ is a continuous, linear map from $X[t]$ into a locally convex space $Y[s]$, show that $\phi(B)$ is s-bounded in Y whenever B is t-bounded in X.

(d)   Let $(E, \|\cdot\|)$ be a normed space. Refer to the Banach–
      Steinhaus theorem to prove: A subset of E is norm
      bounded iff it is $\sigma(E, E')$-bounded (Exercises 3.3, problem
      4).

4.   The Hahn–Banach Theorem.
   Reflexive Banach Spaces

   The topology of any locally convex space can be defined by means
of a family of seminorms. This makes locally convex spaces convenient
to work with. But these spaces are convenient for another reason. We
shall see that an analog of the Hahn–Banach theorem is true for locally
convex spaces. One consequence of this result will enable us to prove
that a Banach space whose unit ball is compact for the weak topology
is necessarily reflexive. The theorem also has other applications and we
shall explore some of these later.
   Recall that if X is a vector space over K, a real-valued, real linear
functional on X is a map f from X into R such that $f(x + y) = f(x) +
f(y)$ for all x, y in X, and $f(\alpha x) = \alpha f(x)$ for all x in X and all α in R.

   Theorem 1.   Let $X[t]$ be a locally convex space, let C be a closed,
convex subset of X, and let x_0 be any point of X that is not in C. Then
there is a t-continuous, real-valued, real linear functional f on X such
that $f(x_0) > \sup\{f(x)\,|\,x \text{ in } C\}$.

   Proof.   There is a family of seminorms $\{p_\gamma\}$ on X such that
$t = t(\{p_\gamma\})$. Also, there is a t-neighborhood U of zero in X such that
$x_0 + U$ is disjoint from C. We may suppose that there is a finite set p_1,
$\ldots$, p_n in $\{p_\gamma\}$ and positive numbers ε_1, $\ldots$, ε_n, such that $U = \{x$ in
$X\,|\,p_i(x) < \varepsilon_i$ for $1 \le i \le n\}$. Let $V = \{x$ in $X\,|\,p_i(x) < \varepsilon_i/2$ for $1 \le i \le n\}$,
let $\tilde{C} = \bigcup \{x + V\,|\,x \text{ in } C\}$, and note that $\tilde{C}$ is a t-open set that contains
C but does not contain x_0. We shall now show that there is no loss in
generality in assuming that $\tilde{C}$ is a convex set that has zero in its
interior.
   Take u, v in $\tilde{C}$ and a real number α with $0 \le \alpha \le 1$. Then
$u = x + u_1$, $v = y + v_1$, where x, y are in C and u_1, v_1 are in V. Clearly,

$$\alpha u + (1 - \alpha)v = \{\alpha x + (1 - \alpha)y\} + \{\alpha u_1 + (1 - \alpha)v_1\}.$$

Since the first term on the right is in C and the second is in V their sum is in $\tilde{C}$ and so $\tilde{C}$ is a convex set. Now let z be any point of the t-open set $\tilde{C}$. Clearly $\tilde{C} - z$ has zero in its interior and if we can find a t-continuous, real-valued, real linear functional f on X such that $f(x_0 - z) > \sup\{f(y)|y \text{ in } \tilde{C} - z\}$ then we shall have proved the theorem. So we can, and now do, assume that $\tilde{C}$ is a convex set that contains zero in its interior.

Let p be the gauge function of $\tilde{C}$ and note that since $\tilde{C}$ is a neighborhood of zero, p is real valued (Section 3, Lemma 2). Regard X as a vector space over R, define $f(\lambda x_0)$ to be $\lambda p(x_0)$ for all real scalars λ, and note that $f(x) \le p(x)$ for all x in the subspace $\{\lambda x_0 | \lambda \text{ in } R\}$ of X. We now refer to part (a) of the proof of the Hahn–Banach theorem (Section 3.2, Theorem 1). It is shown there that we can extend f to a real-valued, real linear functional F on X in such a way that $F(x) \le p(x)$ for all x. Clearly

$$F(x_0) = p(x_0) > 1 \ge \sup\{p(x)|x \text{ in } \tilde{C}\}$$
$$\ge \sup\{p(x)|x \text{ in } C\} \ge \sup\{F(x)|x \text{ in } C\}.$$

Hence we shall be finished once we have shown that F is t-continuous on X.

Since $F(x) \le p(x)$ on X we must have $F(-x) \le p(-x)$ or $F(x) \ge -p(-x)$ for all x in X. Hence for all x in X, $-p(-x) \le F(x) \le p(x)$. Now $\tilde{C}$ contains a balanced, t-neighborhood U of zero (Section 3, Lemma 2). Clearly, $p(-x) = p(x)$ for x in U and so p, and hence F, is bounded by one on U. It follows that F is t-continuous on X.

Corollary 1. Let $X[t]$ be a locally convex space, let C be a balanced, closed, convex subset of X, and let x_0 be a point of X that is not in C. Then there is a continuous, linear functional h on X such that $|h(x_0)| > \sup\{|h(x)| \,|\, x \text{ in } C\}$.

Proof. Regard X as a vector space over R and use Theorem 1 to find a continuous, real-valued, real linear functional f on X such that $f(x_0) > \sup\{f(x)|x \text{ in } C\}$. There is a number α that is strictly less than $f(x_0)$ and strictly greater than the supremum. Since C is balanced we can assume that $\alpha = 1$. Define $h(x) \equiv f(x) - if(ix)$ for all x in X. By part (b) of the proof of the Hahn–Banach theorem (Section 3.2, Theorem 1), h is a continuous, linear functional on the complex vector space X. Also, $|h(x_0)| \ge |f(x_0)| > 1$. If x is any element of C we have

$$h(x) = \rho e^{i\theta}, \qquad |h(x)| = \rho = e^{-i\theta}h(x) = h(e^{-i\theta}x).$$

But ρ is a real number and so $h(e^{-i\theta}x) = \rho$ means $h(e^{-i\theta}x) = f(e^{-i\theta}x)$. Since $e^{-i\theta}x$ is in C because C is balanced and f is bounded by 1 on C, $\rho \leq 1$.

Let us return now to the study of the weak topology of a normed space. Theorem 1 has some interesting implications in this situation. We shall continue numbering our results as we did in Section 1.

(7)   *A convex subset of a normed space $(E, \|\cdot\|)$ is closed for the norm topology iff it is closed for the weak topology.*

We need only prove that a convex, norm closed subset of E is closed for the weak topology. Let C be such a set and suppose that $x_0 \in E$ is not in C. By Theorem 1 there is a real-valued, real linear functional f on E that is norm continuous and satisfies $f(x_0) > \alpha > \sup\{f(x)\,|\,x \text{ in } C\}$; here α is some real number. At this point we distinguish two cases:

(a) If E is a vector space over R, then $f \in E'$ and so f is $\sigma(E, E')$-continuous on E. But then $\{x \text{ in } E \,|\, f(x) > \alpha\}$ is a weak neighborhood of x_0 that is disjoint from C. Since x_0 was any point of E that is not in C we conclude that C is $\sigma(E, E')$-closed.

(b) If E is a vector space over C we could reason as we did in (a) once we show that f is weakly continuous on E. We know (see part (b) of the proof of the Hahn–Banach theorem) that $h(x) \equiv f(x) - if(ix)$ is in E'. It follows that $h(x)$ is $\sigma(E, E')$-continuous on E. Clearly $\overline{h(x)} = f(x) + if(ix)$ is also $\sigma(E, E')$-continuous on E; it is the composition of $h(x)$ and the map that takes each $z \in C$ to $\bar{z} \in C$. But since $f(x) = \frac{1}{2}[h(x) + \overline{h(x)}]$ we see that f is $\sigma(E, E')$-continuous on E.

(8)   *In any infinite dimensional normed space $(E, \|\cdot\|)$ the norm closed set $S = \{x \text{ in } E \,|\, \|x\| = 1\}$ is dense in the unit ball of E for the weak topology.*

Let $\mathscr{B}$ be the unit ball of E, $\mathscr{B} = \{x \,|\, \|x\| \leq 1\}$, and note that $\mathscr{B}$ is $\sigma(E, E')$-closed by (7). Since $S \subset \mathscr{B}$ we need only show that any point of E whose norm is less than one is in the weak closure of S. Choose $x_0 \in E$, $\|x_0\| < 1$, and let V be a weak neighborhood of x_0. Then V contains $x_0 + U$, where U is of the form $\{x \text{ in } E \,|\, |f_i(x)| < \varepsilon_i \text{ for } 1 \leq i \leq n\}$. So

$$V \supset \{y \,|\, y \in x_0 + U\}$$

$$= \{y \,|\, y - x_0 \in U\} = \{y \,|\, |f_i(y - x_0)| < \varepsilon_i \text{ for } 1 \leq i \leq n\}.$$

Since E is infinite dimensional we can choose z in $\bigcap_{i=1}^{n} N(f_i)$, $z \neq 0$; here $N(f_i) = \{x \mid f_i(x) = 0\}$. Note that, for any scalar α, αz is in $\bigcap_{i=1}^{n} N(f_i)$ and so $x_0 + \alpha z$ is in V. Since the map $\alpha \rightarrow \alpha z$ is continuous from K to E, the map $\alpha z \rightarrow \alpha z + x_0$ is continuous from E to E, and the map $\alpha z + x_0 \rightarrow \|\alpha z + x_0\|$ is continuous from E to K, it follows that $\alpha \rightarrow \|\alpha z + x_0\|$ is a continuous map from K to K. We are trying to show that S is $\sigma(E, E')$-dense in $\mathscr{B}$. We chose x_0, $\|x_0\| < 1$, and an arbitrary weak neighborhood V of x_0. What we must do is show that $V \cap S \neq \varnothing$. When $\alpha = 0$, $\|\alpha z + x_0\| = \|x_0\| < 1$. Writing $\alpha z = \alpha z + x_0 - x_0$ we see that $|\alpha| \|z\| - \|x_0\| \leq \|\alpha z + x_0\|$ and so $\|\alpha z + x_0\|$ tends to infinity as α tends to infinity. By continuity, then, there is some number α_0 such that $\|\alpha_0 z + x_0\| = 1$, and, since $\alpha_0 z + x_0$ is always in V, $\alpha_0 z + x_0$ is in $S \cap V$. This proves (8).

The unit ball of E'' (call it $\mathscr{B}''$) is weak* closed by Alaoglu's theorem (Theorem 1 of Section 1). Let $\mathscr{B}$ be the unit ball of E, regard E as a subspace of E'', and let cl $\mathscr{B}$ denote the $\sigma(E'', E')$-closure of $\mathscr{B}$. We know cl $\mathscr{B} \subset \mathscr{B}''$. Suppose ϕ_0 is in $\mathscr{B}''$ but is not in cl $\mathscr{B}$. Since cl $\mathscr{B}$ is a balanced, closed, convex subset of $E''[\sigma(E'', E')]$ there is, by Corollary 1, a continuous, linear functional f on this space such that $|f(\phi_0)| > \sup\{|f(x)| \mid x \text{ in cl } \mathscr{B}\}$. But, by (4), $f \in E'$. So

$$|f(\phi_0)| \leq \|f\| \|\phi_0\| \leq \|f\| = \sup\{|f(x)| \mid x \text{ in } \mathscr{B}\}$$
$$\leq \sup\{|f(x)| \mid x \text{ in cl } \mathscr{B}\},$$

which is a contradiction. We have proved:

(9) (H. Goldstine) *The unit ball of a normed space $(E, \|\cdot\|)$ is dense in the unit ball of E'' for the weak* topology.*

Theorem 2. A Banach space is reflexive iff its unit ball is compact for the weak topology.

Proof. We have already proved the necessity of this condition (Section 1, Corollary 1 to Theorem 1). Let $(B, \|\cdot\|)$ be a Banach space whose unit ball $\mathscr{B}$ is compact for the weak topology. Since $\sigma(B, B')$ is just the restriction to $B \subset B''$ of the topology $\sigma(B'', B')$, $\mathscr{B}$ is $\sigma(B'', B')$-compact in B''. But by (9) this says that the unit ball of B coincides with the unit ball of B'', and hence that $B = B''$.

CHAPTER 5

More about Weak Topologies

1. Dual Spaces and the Krien–Milman Theorem

At the beginning of Chapter 4 we asked whether every Banach space is equivalent to a dual space. We are now in a position to answer this question. First, let us be more precise about what we mean by a "dual space." We shall say that a Banach space B is a dual space if there is a Banach space B_1 such that B and B_1' are equivalent (Section 3.1, Definition 2). Since a normed space and its completion have the same dual (Exercises 3.3, problem 1b), the assumption that B_1 is a Banach space involves no loss in generality. As a matter of fact, the assumption that B and B' are equivalent, rather than just topologically isomorphic, also involves no loss in generality (see problem 1).

Lemma 1. If the Banach space $(B, \|\cdot\|)$ is a dual space, then there is a Hausdorff, locally convex topology t on B, compatible with the vector space structure of B (Section 4.3, Definition 1), such that the unit ball of $(B, \|\cdot\|)$ is t-compact.

Proof. By hypothesis there is a Banach space B_1 and an equivalence u from B onto B_1'. The adjoint of u (Exercises 4.1, problem 4) u^* is a linear map from B_1'' onto B', and so $Q \equiv u^*(B_1)$ is a linear subspace of B'. Each f in Q defines a seminorm on B (recall that $p_f(x) = |f(x)|$ for all x in B). The family of all these seminorms defines a locally convex topology on B, compatible with the vector space structure of B (Section 4.3, Lemma 1), which we shall denote by $\sigma(B, Q)$. We will now show that u is a homeomorphism from $B[\sigma(B, Q)]$ onto $B_1'[\sigma(B_1', B_1)]$.

Since u^* is an equivalence (Exercises 4.1, problem 4b), a typical $\sigma(B, Q)$-neighborhood of zero is formed as follows: Take a finite set y_1, $y_2, \ldots, y_n$ in B_1, the same number of positive real numbers $\varepsilon_1, \varepsilon_2, \ldots,$ ε_n, and let $V = \{x \in B \mid |u^*(y_i)x| \leq \varepsilon_i \text{ for } 1 \leq i \leq n\}$. Clearly

$$u(V) = \{f \in B_1' \mid f = u(x) \text{ for some } x \in V\}$$

$$= \{f \in B_1' \mid f = u(x) \text{ and } |u^*(y_i)x| \leq \varepsilon_i \text{ for } 1 \leq i \leq n\}$$

$$= \{f \in B_1' \mid |f(y_i)| \leq \varepsilon_i \text{ for } 1 \leq i \leq n\}$$

(see the definition of u^*). This last set is a typical $\sigma(B_1', B_1)$-neighborhood of zero in B_1'. Thus u is a homeomorphism between these two spaces (see the discussion in Section 4.3 just after Definition 1). In particular, we see that $\sigma(B, Q)$ is a Hausdorff topology.

Now set $t = \sigma(B, Q)$. It follows immediately that, since u maps the unit ball of B onto the unit ball of B_1', the unit ball of B is t-compact (Section 4.4, Theorem 1).

Krein and Milman [17] were the first to prove that the compact subsets of a Hausdorff, locally convex space have a useful geometric property. Their remarkable theorem will enable us to settle the question under discussion here, and this is only one of its many applications.

Definition 1.   Let X be a vector space over K, let A be a nonempty subset of X, and let B be a subset of A. If the conditions x, y in A, α a real number such that $0 < \alpha < 1$, and $\alpha x + (1 - \alpha)y$ in B imply that both x and y are in B, then we shall say that B is an extreme subset of

A. An extreme subset of A that consists of a single point will be called an extreme point of A.

For example, if A is the solid unit square in R^2 (i.e., $A = \{(x, y) \mid 0 \le x \le 1, \, 0 \le y \le 1\}$) then the boundary of A is an extreme subset of A, and the vertices of the square are extreme points of A. If A is the unit disk in the complex plane then each point of the unit circle is an extreme point of A.

Theorem 1 (Krein–Milman Theorem). Every compact subset of a Hausdorff, locally convex space has extreme points. Furthermore, the closed, convex hull of the extreme points of any such set is equal to the closed, convex hull of the set itself.

Proof. Let $X[t]$ be a Hausdorff, locally convex space and let A be a compact subset of X. Let $\mathscr{P}$ be the family of all closed, extreme subsets of A and notice that $\mathscr{P}$ is not the empty family since it contains A. Partially order $\mathscr{P}$ by inclusion, let $\mathscr{C}$ be a chain in this partially ordered set, and note that, since A is compact, the set $T = \bigcap\{S \mid S \in \mathscr{C}\}$ is nonempty. It is easy to see that T is a lower bound for $\mathscr{C}$ and hence by Zorn's lemma $\mathscr{P}$ contains a minimal element A_0.

Suppose that A_0 contains two distinct points x_1, x_2. Choose a continuous, real-valued, real linear functional f on $X[t]$ such that $f(x_1) \ne f(x_2)$ (Section 4.4, Theorem 1). Let $A_1 = \{x \in A_0 \mid f$ attains its maximum on A_0 at $x\}$. Clearly A_1 is a closed, proper subset of A_0. If we can show that A_1 is an extreme subset of A, then we shall have reached a contradiction. With this in mind we suppose that x, y are in A, that α is a real number such that $0 < \alpha < 1$, and that $\alpha x + (1 - \alpha)y$ is in A_1. This last condition says that f attains its maximum over A_0 at $\alpha x + (1 - \alpha)y$. But since f is linear this means that f attains its maximum over A_0 at both x and y; i.e., both x and y are in A_1. Thus A has extreme points.

Let B be the set of all extreme points of A, and let c.c.(B), c.c.(A) denote the closed, convex hulls of B and A, respectively. To prove that these two sets are equal it is sufficient to prove that A is contained in c.c.(B). Suppose that the point $z \in A$ is not in c.c.(B). We can (Section 4.4, Theorem 1) find a continuous, real-valued, real linear functional g on $X[t]$ such that $g(z) > \sup\{g(x) \mid x$ in c.c.$(B)\}$. Let $A_2 = \{y \in A \mid g$ attains its maximum on A at $y\}$. Since A_2 is a nonempty compact set, it has an extreme point w. But because of the way g was chosen, w could not be an extreme point of A. Thus there are distinct points w_1, w_2 of A and a real number α, $0 < \alpha < 1$, such that $w = \alpha w_1 + (1 - \alpha)w_2$. But since g is linear and attains its maximum on A at w, it attains this maximum at each of the points w_1, w_2; i.e., both w_1 and w_2 are in A_2. This contradicts the fact that w is an extreme point of A_2.

Combining the Krein–Milman theorem with Lemma 1 we see that a Banach space whose unit ball has no extreme points could not be a dual space. Such spaces do exist. Consider, for instance, $(L_1[0, 1], \|\cdot\|_1)$. Given f in this space with $\|f\|_1 = 1$, we can choose a real number c such that $\int_0^c |f(t)|\ dt = 2^{-1}$. Having done that we define f_1 and f_2 as follows: $f_1(t) = 2f(t)$ for $0 \leq t < c$, $f_1(t) = 0$ for $c \leq t$; $f_2(t) = 0$ for $0 \leq t < c$ and $f_2(t) = 2f(t)$ for $c \leq t \leq 1$. Clearly f_1, f_2 are in the unit ball of $L_1, f_1 \neq f_2$, and $f = 2^{-1}f_1 + 2^{-1}f_2$.

EXERCISES 1

1.   Let B be a Banach space and suppose that there is a Banach space B_1 such that B and B_1' are topologically isomorphic. Show that there is a Banach space B_2 such that B and B_2' are equivalent.

2.   Let X be a vector space over K, let A be a balanced subset of X, and let x be an extreme point of A. Show that for any $\sigma \in K$, $|\sigma| = 1$, σx is also an extreme point of A.

3.   (a)   Show that the extreme points of the unit ball of l_∞ are the points $\{x_n\} \in l_\infty$ such that $|x_n| = 1$ for all n.
     (b)   Show that the extreme points of the unit ball of l_1 are the points $\sigma e_n, n = 1, 2, \ldots$, where $|\sigma| = 1$, and e_n is the sequence with one in the nth place and zeros elsewhere.
     (c)   Show that $(c_0, \|\cdot\|_\infty)$ is not a dual space.

4.   (a)   Let B be an infinite dimensional Banach space whose unit ball has only a finite number of extreme points. Show that B is not a dual space.
     (b)   Show that a point f in $(C[0, 1], \|\cdot\|_\infty)$ is an extreme point of the unit ball of this space iff $|f(t)| = 1$ for all t.

(c) Show that the Banach space of real-valued, continuous functions on $[0, 1]$ is not a dual space.

2. The Eberlein–Smulian Theorem

We have seen that the unit ball in the dual of any normed space, is weak* compact (Section 4.1, Theorem 1). If the normed space is separable, then any sequence of points in this ball has a weak* convergent subsequence (Exercises 4.1, problem 5). Let us take a look at the dual of the nonseparable space $(l_\infty, \|\cdot\|_\infty)$. For each n define $u_n \in l'_\infty$ as follows: $u_n(\{x_k\}) = x_n$ for every $\{x_k\} \in l_\infty$. Clearly $u_1, u_2, u_3, \ldots$ are all in the unit ball of l'_∞. Let $\{u_{n_j}\}$ be any subsequence of $\{u_n\}$. We shall show that $\{u_{n_j}\}$ is not weak* convergent. All we have to do is to find $\{x_k\} \in l_\infty$ for which $\lim_{j\to\infty} u_{n_j}(\{x_k\})$ does not exist. Define a sequence as follows: If $k = n_j$ let $x_k = 2^{-1}[1 + (-1)^j]$; if $k \neq n_j$ for all j, let $x_k = 0$. Then $\{x_k\}$ is in l_∞ and $u_{n_j}(\{x_k\}) = 2^{-1}[1 + (-1)^j]$. Clearly $\lim u_{n_j}(\{x_k\})$ does not exist and we have proved:

(10) *A sequence of points of a weak* compact set need not have a weak* convergent subsequence.*

Our result (10) is in striking contrast to the situation in a metric space. We know that a subset A of a metric space is compact iff every sequence of points of A has a subsequence that converges to a point of A. Also, A is compact iff every sequence of points of A has an adherent point in A. Now let E be a normed space. Which of the above equivalences is true for subsets of the topological space $E[\sigma(E, E')]$? This question will occupy us for the remainder of this chapter. Our first theorem, which will take a while to prove, is a generalization of a result of Smulian [see 16, p. 311].

Theorem 1. Let $(E, \|\cdot\|)$ be a normed space and let A be a subset of E that is $\sigma(E, E')$-compact. Then every sequence of points of A has a subsequence that is $\sigma(E, E')$-convergent to a point of A.

(11) *Let $(E, \|\cdot\|)$ be a separable, normed space. Then E' contains a countable, weak* dense set.*

Let $\{x_n\}$ be a countable, total subset of E (Exercises 4.1, problem 3) such that $\|x_n\| = 1$ for all n. Define a metric d on E' as follows: For each pair $(f, g) \in E' \times E'$ let $d(f, g) = \sum_{n=1}^{\infty} 2^{-n}|(f-g)(x_n)|$. Let $\mathscr{B}'$ be the unit ball of E' and consider the identity map I from $\mathscr{B}'$ with the topology induced on it by $\sigma(E', E)$ onto $\mathscr{B}'$ with the topology induced on it by d. We shall show that I is a homeomorphism.

If $f_0 \in \mathscr{B}'$ then a d-neighborhood of f_0 is

$$\{f \in \mathscr{B}' \,|\, d(f, f_0) < \varepsilon\} = \left\{ f \in \mathscr{B}' \,\middle|\, \sum_{n=1}^{\infty} 2^{-n}|(f - f_n)x_n| < \varepsilon \right\}.$$

Since $\mathscr{B}'$ is norm bounded we can choose m so that $\sum_{n+1}^{\infty} 2^{-n}|(f-g)x_n| < \varepsilon/2$ for all f, g in $\mathscr{B}'$. Thus I^{-1} of our d-neighborhood of f_0 contains the $\sigma(E', E)$-neighborhood $\{f \in \mathscr{B}' \,|\, |(f - f_0)x_n| < 2^n(2m)^{-1}\varepsilon$ for $n = 1, 2, \ldots, m\}$ of f_0, i.e., I is continuous. But since $\mathscr{B}'$ is $\sigma(E', E)$-compact, I is a homeomorphism [21, Proposition 5, p. 159].

We have just shown that $\mathscr{B}'$, with the topology induced on it by $\sigma(E', E)$, is a compact metric space. It follows [21, Proposition 13, p. 163] that $\mathscr{B}'$ contains a countable, $\sigma(E', E)$-dense set. But $E' = \bigcup_{n=1}^{\infty} n\mathscr{B}'$ and so E' also contains such a set.

(12) *Let $(E, \|\cdot\|)$ be a separable, normed space and let A be any $\sigma(E, E')$-compact subset of E. Then the topology induced on A by $\sigma(E, E')$ is metrizable.*

Let $\{f_n\}$ be any weak* dense sequence in E'. We can define a metric d on E by letting $d(x, y) = \sum_{n=1}^{\infty} 2^{-n}|f_n(x - y)|$ for all x, y in E. Since, by the Banach–Steinhaus theorem, A is bounded, the argument used to prove (11) shows that the topology induced on A by $\sigma(E, E')$ coincides with the topology induced on A by d.

(13) *Let $(E, \|\cdot\|)$ be a normed space and let H be a closed, linear subspace of E. Then the dual of H is equivalent to E'/H.*

Recall that $H^{\perp} = \{f \in E' \,|\, f \text{ vanishes on } H\}$. The quotient norm was defined in Section 2.4, Definition 1. For $f \in H'$ consider the set $F = \{g \in E' \,|\, f = g \text{ on } H\}$. We can identify this set with an element of $E'/H^{\perp}$, which we shall call $\phi(f)$. Thus we have a map ϕ from H' onto $E'/H^{\perp}$, and clearly ϕ is an isomorphism. Now $\|\phi(f)\| = \inf\{\|g\| \,|\, g \in F\}$ and so $\|\phi(f)\| \geq \|f\|$. But, by the Hahn–Banach theorem, there is a $g \in F$ such that $\|g\| = \|f\|$. Hence $\|\phi(f)\| = \|f\|$ for all $f \in H'$.

This result has an immediate corollary:

(14) *Let $(E, \|\cdot\|)$ be a normed space and let H be a closed, linear*

subspace of E. Then the restriction to H of the topology $\sigma(E, E')$
coincides with the topology $\sigma(H, H')$.

We can now prove Theorem 1. Let $\{a_n\}$ be any sequence of points of
A and let H be the closed, linear subspace of E generated by $\{a_n\}$. H is
$\sigma(E, E')$-closed (Section 4.1, (3)) and so $H \cap A$ is $\sigma(H, H')$-compact in
H (we are using (14)). But H is separable, so, by (12), $\{a_n\}$ has a
$\sigma(E, E')$-convergent subsequence.

W. F. Eberlein [5] was the first to show that if $(B, \|\cdot\|)$ is a Banach
space, a subset A of B is $\sigma(B, B')$-compact iff every sequence of points
of A has a $\sigma(B, B')$-adherent point in A. We shall present his proof
below. The following result, which is true only for Banach spaces (see
(16)), plays a crucial role in this proof.

(15)   *Let $(B, \|\cdot\|)$ be a Banach space. A norm closed, hyperplane in B' is
weak* closed iff its intersection with the unit ball of B' is weak*
closed.*

Recall that a hyperplane in B' is a linear subspace that has codimension one in B'. The necessity of our condition is obvious. Let H be
a norm closed hyperplane in B', let $\mathscr{B}'$ be the unit ball of B', and let
$\mathrm{cl}(H \cap \mathscr{B}')$ denote the weak* closure of $H \cap \mathscr{B}'$. We want to show
that $\mathrm{cl}(H \cap \mathscr{B}') = H \cap \mathscr{B}'$ implies H is weak* closed in B'. The proof
will be given in stages:

(a) H is either weak* closed or weak* dense in B'.

We know that H is the null space of some element $\phi \in B''$ (Exercises 1.3, problem 4, and Section 1.3, Lemma 1). If $\phi \notin B$, then H is
weak* dense in B' (Exercises 4.3, problem 1c).

(b) If H is weak* dense in B', then $B \oplus H^\perp$ is norm closed in B''.
Consequently, there is an $\alpha > 0$ such that $\alpha\|x\| \le \|x + y\|$ for all $x \in B$,
all $y \in H^\perp$.

Let ϕ be an element of B'' whose null space is H. Then $H^\perp = \{\lambda\phi \,|\, \lambda$
in $K\}$ and clearly $B \cap H^\perp = \{0\}$. Let $\{z_n\}$ be a sequence of points of
$B \oplus H^\perp$ that is norm convergent to $z_0 \in B''$. Since B is a Banach space
it is closed in B'' and so there is a continuous, linear functional ψ on B''
such that $\psi(\phi) \ne 0$, $\psi(x) = 0$ for all $x \in B$ (by the Hahn–Banach
theorem). Now for each n, $z_n = x_n + \lambda_n \phi$ where $x_n \in B$, $\lambda_n \in K$. Applying ψ to $\{x_n + \lambda_n \phi\}$ we see that both $\{x_n\}$ and $\{\lambda_n\}$ converge to, say,
$x_0 \in B$, $\lambda_0 \in K$. Clearly $z_0 = x_0 + \lambda_0 \phi$.

The second statement follows from the first and Section 2.5, Corollary 1 to Theorem 1.

We shall say that $\text{cl}(H \cap \mathscr{B}')$ contains a ball iff there is an $\varepsilon > 0$ such that $\{f \in B' \mid \|f\| \leq \varepsilon\} \subset \text{cl}(H \cap \mathscr{B}')$.

(c) If $\text{cl}(H \cap \mathscr{B}')$ does not contain a ball then, for any $\varepsilon > 0$, there is an x_ε in B such that $\sup\{|f(x_\varepsilon)| \mid f \text{ in } H \cap \mathscr{B}'\} \leq \varepsilon \|x_\varepsilon\|$.

Suppose that this is false. Then for some $\delta > 0$ we have $\sup\{|f(x)| \mid f \in H \cap \mathscr{B}'\} > \delta \|x\|$ for all $x \in B$. Now the ball $\mathscr{B}'_\delta$ is not contained in $\text{cl}(H \cap \mathscr{B}')$ by hypothesis. Thus there is a point $f_0 \in \mathscr{B}'_\delta$ that is not in $\text{cl}(H \cap \mathscr{B}')$. Since this weak* closure is balanced and convex, there is a point $x_0 \in B$ such that $x_0(f_0) > \sup\{|f(x_0)| \mid f \in H \cap \mathscr{B}'\} > \delta \|x_0\|$. Thus

$$\|x_0\| \, \|f_0\| \geq x_0(f_0) > \delta \|x_0\|$$

implies $\|f_0\| > \delta$, a contradiction.

(d) H is weak* dense in B' iff $\text{cl}(H \cap \mathscr{B}')$ contains a ball. If $\text{cl}(H \cap \mathscr{B}')$ contains a ball then, since the weak* closure of H contains $\bigcup_{n=1}^\infty n \, \text{cl}(H \cap \mathscr{B}')$, H is weak* dense in B'.

Assume that H is weak* dense in B'. We want to prove that $\text{cl}(H \cap \mathscr{B}')$ contains a ball. Suppose that it does not. Then, for any $\varepsilon > 0$, we can find x_ε in B such that $\sup\{|f(x_\varepsilon)| \mid f \text{ in } H \cap \mathscr{B}'\} \leq \varepsilon \|x_\varepsilon\|$ by (c). Regard x_ε as a linear functional on H and extend it, without changing its norm, to a linear functional x^* on B'. Clearly $x^* = x_\varepsilon + y$, $y \in H^\perp$. Also,

$$\|x^*\| = \sup\{|x^*(f)| \mid f \in \mathscr{B}'\} = \sup\{|x_\varepsilon(f)| \mid f \in H \cap \mathscr{B}'\} \leq \varepsilon \|x_\varepsilon\|$$

because the norm of x^* is equal to the norm of x_ε on H. So $\|x_\varepsilon + y\| \leq \varepsilon \|x_\varepsilon\|$. Now by (b) we have $\alpha \|x_\varepsilon\| \leq \|x_\varepsilon + y\|$, hence $\alpha \|x_\varepsilon\| \leq \varepsilon \|x_\varepsilon\|$. Since $\varepsilon > 0$ is arbitrary, this last inequality is impossible.

Finally, if $H \cap \mathscr{B}'$ is weak* closed then, since H is a proper subspace of B', $H \cap \mathscr{B}'$ could not contain a ball. It follows from (d) and (a) that H is weak* closed in B'.

(16)   *A normed space $(E, \|\cdot\|)$ is a Banach space iff any norm closed, hyperplane in E' whose intersection with the unit ball of E' is weak* closed, is weak* closed.*

To prove (16) we need only show that in the dual of any incomplete normed space there is a norm closed, weak* dense hyperplane whose intersection with the unit ball is weak* closed. Let $(E, \|\cdot\|)$ be an incomplete normed space and let $(\hat{E}, \|\cdot\|)$ be its completion (Exercises 3.4, problem 1a). We know that E and $\hat{E}$ have the same dual space E' (Exercises 3.4, problem 1b). Let $\hat{x} \in \hat{E}$, $\hat{x} \notin E$ and regard $\hat{x}$ as a linear functional on E'. $\hat{x}$ is not $\sigma(E', E)$-continuous on E' (Section 4.1, (4)). So the null space of $\hat{x}$, $N(\hat{x})$, is $\sigma(E', E)$-dense in E' (Exercises 4.3, problem 1c). But, if $\mathscr{B}'$ is the unit ball of E', the set $N(\hat{x}) \cap \mathscr{B}'$ is $\sigma(E', \hat{E})$-compact. Thus the Hausdorff topology $\sigma(E', E)$ must coincide with $\sigma(E', \hat{E})$ on this set [21, Proposition 5, p. 159]. Hence $N(\hat{x}) \cap \mathscr{B}'$ is $\sigma(E', E)$-closed.

Remark 1. In the dual of a Banach space the following is true: A norm closed, linear subspace is weak* closed iff its intersection with the unit ball is weak* closed. I believe that this was first proved by S. Banach. A proof can be found in [16, (6), p. 273].

Theorem 2 (W. F. Eberlein). Let $(B, \|\cdot\|)$ be a Banach space. For any weakly closed subset A of B the two following conditions are equivalent: (a) A is $\sigma(B, B')$-compact. (b) Every sequence of points of A has a $\sigma(B, B')$-adherent point in A.

Proof. Assume (b) and notice that, by the Banach–Steinhaus theorem, this implies A is norm bounded. Thus, by Alaoglu's theorem, the closure of A in $B''[\sigma(B'', B')]$ is $\sigma(B'', B')$-compact. Hence to prove that (b) implies (a) it suffices to show that this closure is contained in B. Let ϕ be an element of B'' that is in the $\sigma(B'', B')$-closure of A. The null space of ϕ, $N(\phi)$, is a norm closed hyperplane in B'. By (15), ϕ will be in B iff $N(\phi) \cap \mathscr{B}'$, where $\mathscr{B}'$ is the unit ball of B', is weak* closed.

To prove that $N(\phi) \cap \mathscr{B}'$ is weak* closed we need only show that if $g \in B'$ is in the $\sigma(B', B)$-closure of this set, then $\phi(g) = 0$. Let $g \in B'$ be in this closure and notice that we can find $x_0 \in A$ such that $g(x_0) = \phi(g)$ because $\{\psi \in B'' \mid |\psi(g) - \phi(g)| < 1/n\}$ contains $x_n \in A$ for $n = 1$, 2, ..., and we can take x_0 to be any $\sigma(B, B')$-adherent point of $\{x_n\}$ that is in A. Given $\varepsilon > 0$, $\{f \in B' \mid |g(x_0) - f(x_0)| < \varepsilon/2\}$ contains a point $f_1 \in N(\phi) \cap \mathscr{B}'$. By the argument just given we can find $x_1 \in A$ such that $\phi(g) = g(x_1)$, $\phi(f_1) = f_1(x_1)$. Now $\{f \in B' \mid |g(x_i) - f(x_i)| < \varepsilon/2$ for $i = 0, 1\}$ contains $f_2 \in N(\phi) \cap \mathscr{B}'$. So now we choose $x_2 \in A$ such that $\phi(g) = g(x_2)$, $\phi(f_1) = f_1(x_2)$, $\phi(f_2) = f_2(x_2)$. Continue in this way. After x_0, x_1, ..., x_k and f_1, f_2, ..., f_k have been chosen, choose $f_{k+1} \in N(\phi) \cap \mathscr{B}'$, which is in $\{f \in B' \mid |g(x_i) - f(x_1)| < \varepsilon/2$ for $i = 0$, 1, 2, ..., $k\}$, and then choose $x_{k+1} \in A$ such that $\phi(g) = g(x_{k+1})$, $\phi(f_j) = f_j(x_{k+1})$ for $j = 1, 2, ..., k + 1$. In this way we generate two sequences $\{x_n\}_{n=0}^{\infty} \subset A$, $\{f_m\}_{m=1}^{\infty} \subset N(\phi) \cap \mathscr{B}'$, which satisfy: (i) $f_m(x_n) = 0$ for $m \le n$ (because $f_m(x_n) = \phi(f_m) = 0$ for $m \le n$); (ii) $g(x_n) = \phi(g)$ for all n; (iii) $|\phi(g) - f_m(x_n)| < \varepsilon/2$ for $n < m$.

Now we are assuming condition (b) and so $\{x_n\} \subset A$ has a $\sigma(B, B')$-adherent point $y \in A$. So for any $\delta > 0$ and any integer M, $\{x \in B \mid |f_m(x) - f_m(y)| < \delta$ for $m = 1, 2, \ldots, M\}$ contains infinitely many points of the sequence $\{x_n\}$. It follows, from (i), that $|f_m(y)| < \delta$ for $m = 1, 2, \ldots, M$, and this implies $f_m(y) = 0$ for all m. The point y is in the $\sigma(B, B')$-closure of $\{x_n\}$ and so (Section 4.4, (7)) it is in the norm closure of the convex hull of $\{x_n\}$ (for the definition of convex hull see Exercises 4.2, problem 1b). So there is a z in this convex hull such that $\|y - z\| < \varepsilon/2$. We must have $z = \sum_{n=1}^{P} a_n x_n$ (Exercises 4.2, problem 1b), where $a_n \geq 0$ for all n and $\sum_{n=1}^{P} a_n = 1$. Set $m = P + 1$ and consider inequality (iii). We have $|\phi(g) - f_m(x_n)| < \varepsilon/2$ for $n < m = P + 1$, and so

$$|\phi(g) - f_{P+1}(x_0)| < \varepsilon/2, \qquad |\phi(g) - f_{P+1}(x_1)| < \varepsilon/2,$$

$$\ldots, \qquad |\phi(g) - f_{P+1}(x_p)| < \varepsilon/2.$$

If we multiply each of these inequalities by the appropriate a_n and add we get $|\phi(g) - f_{P+1}(z)| < \varepsilon/2$. But

$$|\phi(g)| \leq |\phi(g) - f_{P+1}(z)| + |f_{P+1}(z)$$

$$- f_{P+1}(y)| < \varepsilon/2 + \|f_{P+1}\| \|z - y\| < \varepsilon$$

since each $f_n \in \mathscr{B}'$. Now since $\varepsilon > 0$ was arbitrary $\phi(g)$ must be zero and the proof is complete.

Combining Theorems 1 and 2 we obtain a result that is often called the Eberlein–Smulian theorem.

Theorem 3. Let $(B, \|\cdot\|)$ be a Banach space. For any weakly closed subset A of B the three following conditions are equivalent:

(a) A is $\sigma(B, B')$-compact.
(b) Every sequence of points of A has a subsequence that is $\sigma(B, B')$-convergent to a point of A.
(c) Every sequence of points of A has an adherent point in A.

Remark 2. Referring to Theorem 3, a set with property (a) is called a weakly compact set. A set with property (b) is called a sequentially weakly compact set, and one with property (c) is called a countably weakly compact set. For another proof of the Eberlein–Smulian theorem see [26].

EXERCISES 2

1. (a) Let T be a compact space. Show that any sequence of points of T has an adherent point in T.
    (b) Let S be a topological space. Show that the two following conditions on S are equivalent: (i) Every sequence of points of S has an adherent point in S. (ii) Every countable open covering of S has a finite subcovering.

2. Let I be an uncountable set. For each $v \in I$ let $C_v = [0, 1]$ and, in the compact set $\prod \{C_v \mid v \in I\}$, consider $G = \{\{x_v\} \mid x_v = 0$ except for countably many $v \in I\}$. Show that G is not compact and yet every sequence of points of G has a subsequence which converges to a point of G.

3. Let $(E, \|\cdot\|)$ be a normed space and let H be a closed, linear subspace of E. Show that the dual of E/H (Section 2.4, Definition 1) is equivalent to $H^{\perp}$.

4. Show that a Banach space is reflexive iff each of its closed, linear, separable subspaces is reflexive.

CHAPTER **6**

Applications to Analysis

1. Applications to Trigonometric Series

The theory of normed spaces has some nice applications to the study of Fourier series. In order to present these we shall have to recall some of the properties of the L_p-spaces of real variable theory. A detailed treatment of these spaces can be found in either [10] or [21].

One can define an equivalence relation on the set of all complex-valued, measurable functions on $[-\pi, \pi]$ by calling two such functions equal if they differ only on a set of measure zero. The equivalence classes are, by convention, also called functions and these are the elements of our L_p-spaces. For real p, $p \geq 1$, let $L_p = \{f \mid \int |f(x)|^p \, dx < \infty\}$, where we agree that, *throughout this chapter, any integral without limits goes from $-\pi$ to π.* If f is in L_p we define $\|f\|_p$ to be the pth root of $\int |f(x)|^p \, dx$. Each of the spaces $(L_p, \|\cdot\|_p)$ is a Banach space.

If $p > 1$, then the dual of $(L_p, \|\cdot\|_p)$ can be identified with the space $(L_q, \|\cdot\|_q)$, where $p^{-1} + q^{-1} = 1$. This means that if G is any continuous, linear functional on L_p, then there is a unique element g in L_q such that $\|G\| = \|g\|_q$ and $G(f) = \int f(x)\overline{g(x)}\, dx$ for all f in L_p. The dual of $(L_1, \|\cdot\|_1)$ can be identified with the space $(L_\infty, \|\cdot\|_\infty)$, where L_∞ is the space of all essentially bounded functions and, for f in L_∞, $\|f\|_\infty$ is the essential supremum of f. In this connection we also recall the Hölder inequality: If f and g are in L_p and L_q, respectively, where $p^{-1} + q^{-1} = 1$, then $\|fg\|_1 \leq \|f\|_p \|g\|_q$.

It is convenient to identify each f in L_p with its extension, by 2π-periodicity, to all of R; i.e., for any $x \in R$, $f(x) = f(y)$, where y is the unique element of $(-\pi, \pi]$ such that $x - y$ is an integral multiple of 2π. With this convention, for any p, any f in L_p, and any t in R, we can define $f_t(x)$ to be $f(x - t)$ and have $f_t \in L_p$, $\|f_t\|_p = \|f\|_p$. Also, there follows:

Lemma 1.   Let $f \in L_p$, p finite. Then for any s in R, $\lim_{t \to s} \|f_t - f_s\| = 0$.

Proof.   For any $\varepsilon > 0$ we can choose a continuous function g such that $\|f - g\|_p < \varepsilon/4$. Then

$$\|f_t - f_s\|_p \leq \|f_t - g_t\|_p + \|g_t - g_s\|_p + \|g_s - f_s\|_p \leq \varepsilon/2 + \|g_t - g_s\|_p.$$

Since g is continuous this last term can be made as small as we please by choosing $|t - s|$ sufficiently small.

Definition 1.   Let $f \in L_1$. The numbers

$$\hat{f}(n) = (1/2\pi) \int f(x)e^{-inx}\, dx, \qquad n = 0, \pm 1, \pm 2, \ldots,$$

are called the Fourier coefficients of f. The function $\hat{f}$, defined on the integers, is called the Fourier transform of f.

Notice that, since $[-\pi, \pi]$ has finite measure, we have defined the Fourier coefficients, and the Fourier transform, of any L_p-function. We record here, for later use, the following:

Theorem 1 (Riemann–Lebesgue).   If $f \in L_1$, then

$$\lim_{n \to \pm\infty} \hat{f}(n) = 0.$$

Proof.   By definition, $\hat{f}(n) = (1/2\pi) \int f(x)e^{-inx}\, dx$ and, because of our conventions, $-\hat{f}(n) = (1/2\pi) \int f(x) \exp\{-in(x - \pi/n)\}\, dx$. Hence $2\hat{f}(n) = (1/2\pi) \int e^{-inx}\{f(x) - f(x + \pi/n)\}\, dx$ and so

$$2\,|\hat{f}(n)| \le (1/2\pi)\|f - f_{-\pi/n}\|_1 .$$

This last term tends to zero, as $n \to \pm\infty$, by lemma 1.

The following problem is of interest to harmonic analysts. Given a sequence $\{c_n \,|\, n = 0, \pm 1, \pm 2, \ldots\}$, find necessary and sufficient conditions for $\{c_n\}$ to be the Fourier transform of a function in L_p; i.e., we want necessary and sufficient conditions for the existence of a function f, f in L_p for some p, such that $\hat{f}(n) = c_n$ for all n.

It is clear that the partial sums of the series $\sum_{n=-\infty}^{\infty} c_n e^{inx}$ are in L_p for every p. If these partial sums converge for the norm of some L_p space, then the limit is a function in L_p whose Fourier transform is $\{c_n\}$. Thus the convergence for some L_p-norm of our series is a sufficient condition for $\{c_n\}$ to be the Fourier transform of an L_p-function. However, as we shall soon see, we can do much better than this.

At this point we have to recall a familiar topic from the theory of infinite series. If $a_0, a_1, a_2, \ldots$ is any sequence of complex numbers, the Cesaro means of this sequence are the numbers

$$\sigma_1 = a_0, \qquad \sigma_2 = \frac{a_0 + a_1}{2}, \qquad \sigma_3 = \frac{a_0 + a_1 + a_2}{3}, \qquad \ldots .$$

If $\sigma_1, \sigma_2, \sigma_3, \ldots$ converges to, say, σ, then we say that the sequence a_0, $a_1, a_2, \ldots$ is Cesaro convergent to σ. A convergent sequence is Cesaro convergent to its (ordinary) limit, but there are divergent sequences that are Cesaro convergent; $1, 0, 1, 0, 1, 0, \ldots$, for example, is Cesaro convergent to $1/2$. Now consider a series $\sum_{k=-\infty}^{\infty} a_k$. For each $n = 0, 1, 2, \ldots$ let $s_n = \sum_{-n}^{n} a_k$. If this sequence $\{s_n\}$ is Cesaro convergent to, say, s', then we say that the given series is Cesaro convergent to s'.

Recall that we assumed a sequence $\{c_n\}$ was given and we asked for necessary and sufficient conditions for this sequence to be the Fourier transform of an L_p-function. We can give such conditions in terms of the Cesaro means of the series $\sum_{n=-\infty}^{\infty} c_n e^{inx}$ and here, at last, is where we use some functional analysis. Let $\sigma_n(x)$ be the nth Cesaro mean of this series. Then, if $n > |m|$, the mth Fourier coefficient of $\sigma_n(x)$ is $[(n - |m|)/n]c_m$. Hence $\lim_{n \to \infty} (1/2\pi) \int \sigma_n(x) e^{-imx}\, dx = c_m$ for every m. Now suppose that for some $p > 1$ the sequence $\{\sigma_n(x)\}$ is bounded for the L_p-norm; i.e., suppose that for some $p > 1$, $\sup\{\|\sigma_n(x)\|_p \,|\, n = 0, 1, 2, \ldots\} < \infty$. Since $p > 1$, L_p is the dual of the Banach space Lq, where $p^{-1} + q^{-1} = 1$. But then, by Alaoglu's theorem (Section 4.1, Theorem 1), the sequence $\{\sigma_n(x)\}$ has a $\sigma(L_p, L_q)$-adherent point; i.e., there is an $h \in L_p$ such that for any $g \in L_q$, and any $\varepsilon > 0$, $\left| \int [\sigma_n(x)\overline{g(x)} - h(x)\overline{g(x)}]\, dx \right| < \varepsilon$ for infinitely many n. Now e^{imx} is in L_q and so, combining our observations, $\hat{h}(m) = c_m$ for all m. Hence we have proved:

Theorem 2.   If, for some $p > 1$, the Cesaro means of $\sum_{n=-\infty}^{\infty} c_n e^{inx}$ are bounded for the L_p-norm, then the sequence $\{c_n\}$ is the Fourier transform of a function in this space.

Since L_1 is not a dual space (Section 5.1, last paragraph) the proof given above cannot be used for this space. What we do here is simply assume that the Cesaro means converge.

Theorem 3.   If the Cesaro means of the series $\sum_{n=-\infty}^{\infty} c_n e^{inx}$ converge for the L_1-norm, then the sequence $\{c_n\}$ is the Fourier transform of a function in this space.

We shall talk about the converses of these theorems later on. Right now we want to investigate the following question: If the Fourier transform $\hat{f}$ of an L_p-function f is known, can we obtain f and, if so, how? The series $\sum_{n=-\infty}^{\infty} \hat{f}(n)e^{inx}$ and its Cesaro means suggest themselves.

Definition 2.   If $f \in L_1$, then the series $\sum_{n=-\infty}^{\infty} \hat{f}(n)e^{inx}$ is called the Fourier series for f.

We shall need a convenient expression for the Cesaro means of a Fourier series. The nth partial sum of the Fourier series for f (call it $s_n(x; f)$) is

$$\sum_{-n}^{n} \hat{f}(k)e^{ikx} = \sum_{-n}^{n} e^{ikx}\left[(1/2\pi) \int f(y)e^{-iky}\, dy\right]$$

$$= (1/2\pi) \int f(y)\left[\sum_{-n}^{n} e^{ik(x-y)}\right] dy.$$

Hence, if $K_n(x - y)$ denotes the nth Cesaro mean of the series $\sum_{-\infty}^{\infty} e^{ik(x-y)}$, then the nth Cesaro mean of the Fourier series for f (we shall denote it by $\sigma_n(x; f)$ is

(*) $$\sigma_n(x; f) = (1/2\pi) \int f(y)K_n(x - y)\, dy.$$

Lemma 2. Let $K_n(x)$ be the nth Cesaro mean of the series $\sum_{-\infty}^{\infty} e^{ikx}$. Then for $n = 1, 2, 3, \ldots$ we have:

(i) $$K_n(x) = \frac{1}{n}\left(\frac{1 - \cos nx}{1 - \cos x}\right) = \frac{1}{n}\left(\frac{\sin \frac{1}{2}nx}{\sin \frac{1}{2}x}\right)^2;$$

(ii) $K_n(x) \geq 0$ for all x;

(iii) $(1/2\pi) \int K_n(x)\, dx = 1$.

Furthermore, if I is any open interval containing zero, then $\lim_{n\to\infty} \sup\{K_n(x)\,|\,x \in (-\pi, \pi], x \notin I\} = 0$.

Proof. Observe that

$$(n + 1)K_{n+1}(x) - nK_n(x) = \sum_0^n e^{ikx} + \sum_1^n e^{-ikx}.$$

These last two series are geometric and their sums are

$$\frac{1 - e^{i(n+1)x}}{1 - e^{ix}} \quad \text{and} \quad \frac{1 - e^{-i(n+1)x}}{1 - e^{-ix}} - 1,$$

respectively. Adding these we obtain the formula

$$(n + 1)K_{n+1}(x) - nK_n(x) = \frac{\cos nx - \cos(n + 1)x}{1 - \cos x}.$$

Now $K_1(x) \equiv 1$ so, using our formula, we can find $K_2(x)$. Then, knowing $K_2(x)$, we can use our formula again to get $K_3(x)$, and so on. This proves (i) and (ii) follows immediately from (i). To prove (iii) we note that if f is identically equal to 1 then the same is true of each Cesaro mean of the Fourier series for f. Hence, by (*), $1 = (1/2\pi) \int K_n(x - y)\, dy$. Setting $x = 0$ and noting that, from (i), each $K_n(x)$ is an even function, we have (iii).

Finally, if $0 < \delta < \pi$ and $\delta \leq |x| \leq \pi$, then $[\sin(1/2x)]^2 \geq [\sin(1/2\delta)]^2$. So, if I is an open interval containing zero and if $\delta > 0$ is so small that $(-\delta, \delta) \subset I$, then

$$\sup\{K_n(x)\,|\,x \in (-\pi, \pi],\, x \notin I\} \leq \sup\{K_n(x)\,|\,\delta \leq |x| \leq \pi\}$$
$$\leq [\sin(1/2\delta)]^{-2}/n.$$

Clearly the limit, as n tends to infinity, of the last of these is zero.

Theorem 4.   If a function lies in L_p, $1 \leq p < \infty$, then the Cesaro means of its Fourier series converge to this function for the L_p-norm. If the function lies in L_∞, then the Cesaro means of its Fourier series converge to this function for the weak* topology.

Proof.   Let $f \in L_p$, p finite. We want to estimate $\|\sigma_n - f\|_p$, where σ_n is the nth Cesaro mean of the Fourier series for f. Since L_q is the dual of L_p, by the first corollary to the Hahn–Banach theorem (Section 3.2, Theorem 1) we can choose $g \in L_q$ such that $\|g\|_q = 1$ and $\|\sigma_n - f\|_p = \int [\sigma_n(x) - f(x)]\overline{g(x)}\, dx$. Now

$$\left| \int [\sigma_n(x) - f(x)]\overline{g(x)}\, dx \right| = \left| \int \{(1/2\pi) \int [f(x - t) \right.$$
$$\left. - f(x)]K_n(t)\, dt\}\overline{g(x)}\, dx \right|$$

by (*) and (iii) of Lemma 2.

This last integral is

$$\leq (1/2\pi) \int \left| \int [f(x - t) - f(x)]\overline{g(x)}\, dx \right| K_n(t)\, dt,$$

which is, by the Hölder inequality, $\leq (1/2\pi) \int \|f_t - f\|_p K_n(t)\, dt$. So

$$\|\sigma_n - f\|_p \leq (1/2\pi) \int \|f_t - f_p\|K_n(t)\, dt$$

and, if $\delta > 0$ is fixed,

$$\|\sigma_n - f\|_p \leq (1/2\pi) \int_{-\delta}^{\delta} + (1/2\pi) \int_{|t| \geq \delta}$$
$$\leq \sup_{-\delta \leq t \leq \delta} \|f_t - f\|_p + 2\|f\|_p \sup_{|t| \geq \delta} K_n(t).$$

The first of these can be made small by choosing δ sufficiently small (Lemma 1). Once δ is chosen the second term can be made small by choosing n sufficiently large (Lemma 2). This proves the theorem for finite p.

Now suppose that $f \in L_\infty$. For any function $g \in L_1$ we have:

$$\left| \int [\sigma_n(x) - f(x)] \overline{g(x)} \, dx \right|$$

$$\leq \sup_{-\delta \leq t \leq \delta} \left| (1/2\pi) \int [f(x-t) - f(x)] \overline{g(x)} \, dx \right| + 2\|f\|_\infty \sup_{|t| \geq \delta} K_n(t).$$

To prove the theorem it suffices to show that the first term on the right can be made arbitrarily small by choosing δ sufficiently small. But clearly

$$\left| \int [f(x-t) - f(x)] \overline{g(x)} \, dx \right|$$

$$= \left| \int f(y) [\overline{g(y)} - \overline{g(y-t)}] \, dy \right| \leq \|f\|_\infty \|g - g_t\|_1,$$

and the result follows from lemma 1.

Theorem 5. A sequence $\{c_n\}$ is the Fourier transform of a function in L_p, $p > 1$, iff the Cesaro means of the series $\sum c_n e^{inx}$ are uniformly bounded for the L_p-norm. The given sequence is the Fourier transform of a function in L_1 iff the Cesaro means of the series converge for the L_1-norm.

Proof. For $p = 1$ the theorem follows from Theorems 2 and 3. For $1 < p < \infty$, it follows from Theorems 1 and 3. The only thing left to prove is that if $f \in L_\infty$, then the Cesaro means of its Fourier series are uniformly bounded for the L_∞-norm. But this follows from Theorem 3 and the Banach–Steinhaus theorem (Section 3.3, Theorem 1).

Remark. The first person to investigate the Cesaro summability of Fourier series was Fejer [13, 6, p. 8]. The sequence of functions $\{K_n(x)\}$ is called Fejer's kernel.

EXERCISES 1

1. Recall that the space L_2 has an inner product: $\langle f, g \rangle = \int f(x) \overline{g(x)} \, dx$ for all f, g in L_2. Clearly $\langle , \rangle$ is linear in the first

element, $\langle \overline{f, g} \rangle = \langle g, f \rangle$, and $\langle f, f \rangle = \| f \|_2^2$ for all f, g in L_2. A subset H of L_2 is called an orthonormal set if $\langle f, f \rangle = 1$ for each $f \in H$ and $\langle f, g \rangle = 0$ for f, g in H, $f \neq g$.

(a) Show that any orthonormal set in L_2 is countable.

(b) If $\{f_n\}$ is any orthonormal sequence in L_2 and $f \in L_2$, show that $\sum_{n=1}^{\infty} |\langle f, f_n \rangle|^2 \leq \| f \|_2^2$ (Bessel's inequality). (Hint: Consider the inner product of $f - \sum_{n=1}^{m} \langle f, f_n \rangle f_n$ with itself.)

(c) Show that for any orthonormal sequence $\{f_n\}$ in L_2 the following are equivalent:

   (i) $\langle f, f_n \rangle = 0$ for all n implies $f = 0$.

   (ii) For each f in L_2, $f = \sum_{n=1}^{\infty} \langle f, f_n \rangle f_n$, convergence is for the L_2-norm.

   (iii) For each f in L_2, $\| f \|_2^2 = \sum_{n=1}^{\infty} |\langle f, f_n \rangle|^2$ (Parseval's identity).

   (iv) For any f, g in L_2, $\langle f, g \rangle = \sum_{n=1}^{\infty} \langle f, f_n \rangle \langle \overline{g, f_n} \rangle$ (also called Parseval's identity).

   (v)   The linear span of $\{f_n\}$ is dense in L_2 for the norm topology.

(d) An orthonormal sequence in L_2 that has any of the above five properties is called a complete orthonormal sequence.

   (i)   Show that $\{(1/2\pi)^{1/2} e^{ikx} \,|\, k = 0, \pm 1, \pm 2, \ldots\}$ is a complete orthonormal sequence in L_2.

   (ii) Show that the Fourier series of any L_2-function converges to that function for the L_2-norm.

2. Let f be a continuous function on R such that $f(x + 2\pi) = f(x)$ for all x. Let $\sigma_n(x)$ by the nth Cesaro mean of the Fourier series for f. Then we have:

$$|\sigma_n(x) - f(x)| \leq (1/2\pi) \int \left| f(x - t) - f(x) \right| K_n(t)\, dt$$

$$= (1/2\pi) \int_{-\delta}^{\delta} + (1/2\pi) \int_{|t| \geq \delta}$$

(a) Show that

$$|\sigma_n(x) - f(x)| \leq \sup\{| f(x - t) - f(x)| \,|\, -\delta \leq t \leq \delta\}$$
$$+ 2 \max\{| f(x)| \,|\, -\pi \leq x \leq \pi\}$$
$$\sup\{K_n(t) | \,|t| \geq \delta\}.$$

(b) Use (a) to conclude that the Cesaro means of the Fourier series for f converge to f uniformly over $(-\pi, \pi]$.

(c) Use (b) to conclude that the polynomials are dense in $(C[0, 1], \|\cdot\|_\infty)$.

2. Miscellaneous Applications

We are going to present two applications of the Banach–Steinhaus theorem. The first of these is to a problem in the theory of divergent series. Observe that the product

$$
\begin{pmatrix}
1 & 0 & 0 & 0 & 0 & \cdots \\
\frac{1}{2} & \frac{1}{2} & 0 & 0 & 0 & \cdots \\
\frac{1}{3} & \frac{1}{3} & \frac{1}{3} & 0 & 0 & \cdots \\
\frac{1}{4} & \frac{1}{4} & \frac{1}{4} & \frac{1}{4} & 0 & \cdots \\
\vdots & \vdots & \vdots & \vdots & \vdots & \cdots
\end{pmatrix}
\begin{pmatrix}
a_0 \\
a_1 \\
a_2 \\
\vdots
\end{pmatrix}
$$

gives us the Cesaro means of the sequence $a_0, a_1, \ldots$. So the given sequence is Cesaro convergent iff the sequence obtained by multiplying it on the right by the matrix given above is convergent. From this point of view Cesaro convergence is a special case of the following: Let A be a doubly infinite matrix, i.e., $A = (a_{ij})$, where $1 \le i < \infty$, $1 \le j < \infty$. Given $x = (x_1, x_2, \ldots)$, define $A_i(x) = \sum_{k=1}^{\infty} a_{ik} x_k$, $i = 1, 2, \ldots$. We shall say that $x = \{x_n\}$ is A-convergent to the number b if $A_i(x)$ exists for $i = 1, 2, \ldots$ and $\lim_{i \to \infty} A_i(x) = b$.

The problem we want to solve is this: Which doubly infinite matrices A have the property that every convergent sequence is A-convergent to its (ordinary) limit? A matrix that has this property will be called a regular matrix.

Lemma 1. Let $(B, \|\cdot\|)$ be a Banach space and let $\{f_n\}$ be a sequence in B'. There is an $f \in B'$ such that $\lim f_n(x) = f(x)$ for all $x \in B$ iff $\{f_n\}$ is a norm bounded set and $\lim f_n(y) = f(y)$ for all y in some total subset of B.

Proof. If there is an $f \in B'$ such that $\lim f_n(x) = f(x)$ for all x in B, then, since B is a Banach space, $\{f_n\}$ is norm bounded (Section 3.3, Theorem 1).

Assume that $\{f_n\}$ and f satisfy our two conditions. Then clearly $\lim f_n(z) = f(z)$ for all z in the linear span of our total subset. Given

$x \in B$ and $\varepsilon > 0$ choose z in this linear span such that $\|x - z\| < \varepsilon$ (Exercises 4.1, problem 3). Then

$$|f(x) - f_n(x)| \leq \|f\| \|x - z\| + |f(z) - f_n(z)| + \|f_n\| \|x - z\|$$
$$\leq 2M\varepsilon + |f(z) - f_n(z)|,$$

where M is a constant that is greater than $\|f\|$ and $\|f_n\|$ for all n. The last term can be made as small as we please by taking n sufficiently large.

Theorem 1 (Silverman–Toeplitz). The matrix $A = (a_{ij})$ is a regular matrix iff:

(i) There is a number M such that $\sum_{j=1}^{\infty} |a_{ij}| \leq M$ for $i = 1$, 2, ...

(ii) $\lim_i a_{ij} = 0$ for $j = 1, 2, \ldots$.

(iii) $\lim_{i \to \infty} \sum_{j=1}^{\infty} a_{ij} = 1$.

Proof. Clearly c, the vector space of all convergent sequences, is a closed, linear subspace of $(l_\infty, \|\cdot\|_\infty)$. Hence $(c, \|\cdot\|_\infty)$ is a Banach space. Observe that for each $\{x_n\} \in c$, $|\lim x_n| \leq \|\{x_n\}\|_\infty$. So the linear functional $\mathscr{C}$ on c, defined by $\mathscr{C}(\{x_n\}) = \lim x_n$, is continuous on $(c, \|\cdot\|_\infty)$. Let $e_0 = (1, 1, 1, \ldots)$ and, for each n, let $e_n = (0, 0, \ldots, 0, 1, 0, 0, \ldots)$ where the 1 is in the nth place. Clearly each e_n, $n = 0, 1, \ldots$, is in c and $\mathscr{C}(e_0) = 1$, $\mathscr{C}(e_n) = 0$ for $n = 1, 2, \ldots$.

Assume that A has properties (i), (ii), and (iii). Using (i) we can write

$$|A_i(x)| \leq \sum_{j=1}^{\infty} |a_{ij}| \, |x_j| \leq \|\{x_j\}\|_\infty \sum_{j=1}^{\infty} |a_{ij}| \leq M \|\{x_j\}\|_\infty$$

for any $x = \{x_j\} \in c$. So each $A_i \in c'$ and $\|A_i\| \leq M$ for $i = 1, 2, \ldots$. We want to prove that $\lim A_i(x) = \mathscr{C}(x)$ for each $x \in c$. To do that it suffices, by Lemma 1, to show that this holds on a total subset of c. Clearly $e_0, e_1, \ldots$ is such a set. Also, for $j = 1, 2, \ldots, \lim_{i \to \infty} A_i(e_j) = \lim_{i \to \infty} a_{ij} = 0$ (by (ii)) $= \mathscr{C}(e_j)$, and

$$\lim_{i \to \infty} A_i(e_0) = \lim_{i \to \infty} \sum_{j=1}^{\infty} a_{ij} = 1$$

(by (iii)) $= \mathscr{C}(e_0)$. So if A has properties (i), (ii), and (iii), then it is a regular matrix.

Now suppose that A is a regular matrix. Then every convergent sequence is A-convergent to its (ordinary) limit. In particular, e_0, e_1, ... are each A-convergent to their limits. This says that conditions (ii) and (iii) are necessary. We will now show that (i) is necessary. Recall that $A_i(\{x_j\}) = \sum_{j=1}^{\infty} a_{ij} x_j$ and, since A is regular, A_i is a linear functional on c. Fix i and, for each n, let $A_i^{(n)}(x) = \sum_{j=1}^{n} a_{ij} x_j$. Clearly $A_i^{(n)} \in c'$ for $n = 1, 2, \ldots$, and $\lim_{i \to \infty} A_i^{(n)}(x) = A_i(x)$ for every $x \in c$. We conclude that each A_i is in c'.

Now since A is a regular matrix $\lim_{i \to \infty} A_i(x)$ exists for every $x \in c$. Hence, by the Banach–Steinhaus theorem, there is a number M such that $\|A_i\| \le M$ for all i. Choose an integer N and, for $j = 1, 2, \ldots, N$, define $x_j = \bar{a}_{ij} |a_{ij}|^{-1}$ if $a_{ij} \ne 0$, $x_j = 0$ if $a_{ij} = 0$. Then set $x_j = 0$ for all $j > N$. Clearly, $x = \{x_j\}$ is in c, $\|x\|_{\infty} \le 1$, and $A_i(x) = \sum_{j=1}^{N} |a_{ij}| \le \|A_i\| \le M$. Since N is arbitrary we are done.

In 1876 du Bois-Reymond surprised the mathematical community by constructing a continuous function whose Fourier series diverges at a single point [13, Part I, 4, p. 7]. We are going to use the Banach–Steinhaus theorem to prove the existence of such functions. Incidentally, continuous functions whose Fourier series diverge on a dense set (having measure zero) have been constructed [13, Part II, 11, p. 20]. However, to my knowledge, the problem of finding a continuous function whose Fourier series diverges on a set of positive measure, or proving that such functions do not exist, remains open.

Let $C(-\pi, \pi]$ be the space of all continuous functions on R such that $f(x) = f(x + 2\pi)$ for all x. We shall give this space the sup norm (i.e., $\|f\|_{\infty} = \sup\{|f(x)| \mid x \in R\}$), and we note that $(C(-\pi, \pi], \|\cdot\|_{\infty})$ is a Banach space. For any f in this space write the Fourier series of f in the form $\frac{1}{2} a_0 + \sum_{k=1}^{\infty} (a_k \cos kx + b_k \sin kx)$, where

$$a_0 = (1/\pi) \int f(x) \, dx,$$

$a_k = (1/\pi) \int f(x) \cos kx \, dx$, and $b_k = (1/\pi) \int f(x) \sin kx \, dx$, $k = 1$, 2, ... We want to derive an expression for the nth partial sum of this series. To do this we first note that the geometric series

$$\sum_{k=0}^{n} e^{ikx} = \frac{e^{i(n+1)x} - 1}{e^{ix} - 1}$$

and so

$$\frac{1}{2} + \sum_{k=1}^{n} e^{ikx} = \frac{2e^{i(n+1)x} - e^{ix} - 1}{2(e^{ix} - 1)}.$$

Multiply the numerator and the denominator of the right-hand side of this expression by $e^{-ix/2}$ and then equate the real parts of both sides to obtain

$$\frac{1}{2} + \sum_{k=1}^{n} \cos kx = \frac{\sin \frac{1}{2}(2n+1)x}{2 \sin \frac{1}{2}x}.$$

Now let $s_n(x)$ be the nth partial sum of the Fourier series for f. Then

$$s_n(x) = \frac{1}{2}a_0 + \sum_{k=1}^{n} (a_k \cos kx + b_k \sin kx)$$

$$= \frac{1}{2\pi} \int f(y) \, dy + \frac{1}{\pi} \sum_{k=1}^{n} \int f(y)[\cos ky \cos kx$$

$$+ \sin ky \sin kx] \, dy$$

$$= \frac{1}{\pi} \int f(y) \left[\frac{1}{2} + \sum_{k=1}^{n} \cos k(y-x) \right] dy$$

$$= \frac{1}{2\pi} \int f(y) \frac{\sin \frac{1}{2}(2n+1)(y-x)}{\sin \frac{1}{2}(y-x)} \, dy.$$

So, letting

$$D_n(y) = \frac{\sin \frac{1}{2}(2n+1)y}{\sin \frac{1}{2}y}, \qquad n = 0, 1, \ldots$$

we have

$$s_n(x) = \frac{1}{2\pi} \int f(y) D_n(y-x) \, dx.$$

The sequence $\{D_n(y)\}$ is called Dirichlet's kernel.

For each n the map that takes each $f \in C(-\pi, \pi]$ to the nth partial sum of its Fourier series evaluated at zero (i.e., $f \to s_n(0)$) is a linear functional on this space, which we shall denote by u_n. Clearly, $u_n(f) = (1/2\pi) \int f(y) D(y) \, dy$.

Lemma 2.   Each of the functionals u_n, $n = 0, 1, 2, \ldots$, is continuous on $(C(-\pi, \pi], \|\cdot\|_\infty)$. Furthermore, for each n, the norm of u_n is the number $l_n = (1/2\pi) \int |D_n(y)| \, dy$.

Proof.   Fix n and let $\varepsilon > 0$ be given. All we have to do is construct a function $f \in C(-\pi, \pi]$ such that $\|f\|_\infty \leq 1$ and $|u_n(f)| \geq l_n - \varepsilon$.

Let $\text{sgn}(0) = 0$ and, for real $x \neq 0$, let $\text{sgn}(x) = x|x|^{-1}$. Note that $\sin \frac{1}{2}(2n + 1)y = 0$ at the points $y = 2l\pi/(2n + 1)$, $l = 0, \pm 1, \pm 2, \dots$. For $\delta > 0$, $\delta < \pi/(2n + 1)$, define $f_\delta(y) = \text{sgn}(D_n(y))$ for

$$\frac{2l\pi}{2n + 1} + \delta \leq y \leq \frac{2(l + 1)\pi}{2n + 1} - \delta,$$

and define it to be linear for

$$\frac{2l\pi}{2n + 1} - \delta \leq y \leq \frac{2l\pi}{2n + 1} + \delta.$$

Clearly $\|f_\delta\|_\infty = 1$. Also, since $|D_n(y)| = |1 + 2\sum_{k=1}^n \cos ky| \leq 2n + 1$, we see that

$$|u_n(f_\delta)| \geq l_n - \frac{1}{2\pi} \sum_{l=0}^{2n+1} \left\{ \text{integral of } |D_n(y)| \text{ from} \right.$$

$$\left. \frac{2l\pi}{2n + 1} - \delta \text{ to } \frac{2l\pi}{2n + 1} + \delta \right\}$$

$$\geq l_n - \frac{\delta}{\pi} \sum_{l=0}^{2n+1} (2n + 1) = l_n - \frac{\delta}{\pi}(2n + 1)(2n + 2)$$

$$\geq l_n - \varepsilon$$

for δ sufficiently small.

This proves the lemma.

Suppose that the Fourier series of every function in $C(-\pi, \pi]$ is convergent at zero; i.e., suppose that $\lim_{n\to\infty} u_n(f)$ exists for every f in our space. Then, by the Banach–Steinhaus theorem, there is a number M such that $l_n = \|u_n\| \leq M$ for all n. Thus to prove that there is a continuous function whose Fourier series diverges at zero we need only prove that $\lim_{n\to\infty} l_n = \infty$.

In the interval

$$\left[\frac{(4l + 1)\pi}{4n + 2}, \frac{(4l + 3)\pi}{4n + 2} \right]$$

we have $|\sin \frac{1}{2}(2n + 1)y| \geq \sqrt{2}/2$. Hence

$$l_n > \frac{\sqrt{2}}{4\pi} \sum_{l=0}^{2n} \left\{ \text{integral of } \left| \sin \frac{1}{2} y \right|^{-1} \text{ from } \frac{(4l + 1)\pi}{4n + 2} \text{ to } \frac{(4l + 3)\pi}{4n + 2} \right\}.$$

For $y > 0$, $\sin \frac{1}{2}y < \frac{1}{2}y$, and so

$$l_n > \frac{\sqrt{2}}{4\pi} \sum_{l=0}^{2n} \left\{ \text{integral of } 2y^{-1} \text{ from } \frac{(4l+1)\pi}{4n+2} \text{ to } \frac{(4l+3)\pi}{4n+2} \right\}$$

$$> \frac{\sqrt{2}}{\pi} \sum_{l=0}^{2n} \frac{1}{4l+3},$$

and this tends to infinity as $n \to \infty$.

CHAPTER 7

The Theory of Distributions

1. Some Function Spaces. Partitions of Unity

The theory of distributions plays a fundamental role in any modern treatment of partial differential equations (see [7, 11]) and it has important applications to harmonic analysis [14]. That theory will be introduced here. We will need some of the ideas presented in Sections 4.2 and 4.3, and we must begin by discussing various spaces of functions defined on certain subsets of R^n. In all that follows Ω will denote a nonempty, open subset of R^n, where n is arbitrary but, in any discussion, fixed. When we speak of a "function in Ω," without any other qualification, we will always mean a complex-valued function that is defined in Ω and is measurable with respect to Lebesgue measure in Ω. The integral of any function in Ω will always mean the Lebesgue integral, and if no limits are given, we will understand that the integration is over all of Ω. The space of all continuous functions in Ω will be denoted by $C^0(\Omega)$. Note that if f is in this space, then neither $\sup\{|f(x)| \,|\, x \text{ in } \Omega\}$ nor $\int f(x)\,dx$ need be finite.

Definition 1. For each $f \in C^0(\Omega)$ we define the support of f, supp f, to be the closure in Ω of $\{x \,|\, f(x) \neq 0\}$. If supp f is a compact set then we shall say that f has compact support in Ω. The set of all such functions (i.e., $\{f \in C^0(\Omega) \,|\, \text{supp } f \text{ is compact}\}$) will be denoted by $C_0^0(\Omega)$.

If $f \in C_0^0(R^n)$, we can define $\|f\|_\infty$ to be $\sup\{|f(x)| \,|\, x \text{ in } R^n\}$. We leave it to the reader to show that $(C_0^0(R^n), \|\cdot\|_\infty)$ is a Banach space. Also, for any real number $p \geq 1$, we can put a norm $\|\cdot\|_p$ on $C_0^0(R^n)$ as follows: For each f in our space let $\|f\|_p$ be the pth root of $\int |f(x)|^p\,dx$. The completion of the normed space $(C_0^0(R^n), \|\cdot\|_p)$ is the Banach space $(L_p(R^n), \|\cdot\|_p)$ [10].

Definition 2. For each nonnegative integer k let $C^k(\Omega)$ denote the space of all those functions in Ω that have continuous partial derivatives of order up to and including k; by the partial derivatives of order zero we mean, of course, the function itself. We define $C_0^k(\Omega)$ to be $\{f \in C^k(\Omega) \,|\, f \text{ has compact support in } \Omega\}$. The spaces $\bigcap_{k=0}^\infty C^k(\Omega)$ and $\bigcap_{k=0}^\infty C_0^k(\Omega)$ will be denoted by $C^\infty(\Omega)$ and $C_0^\infty(\Omega)$, respectively. An element of $C^\infty(\Omega)$ will be called a C^∞-function in Ω.

It is easy to give examples of C^∞-functions in R^n. The function that takes each $x = (x_1, x_2, \ldots, x_n)$ in R^n to $|x|^2 = x_1^2 + \cdots + x_n^2$ is such a function. Recall that $f(t) = \exp(-t^{-2})$ for $t > 0$ and $= 0$ for $t \leq 0$ is in $C_0^\infty(R)$ [1, pp. 121, 250]. To give an example of a function in $C_0^\infty(R^n)$ we just let $\omega_0(x) = \exp[-(1 - |x|^2)^{-1}]$ for $|x| < 1$ and $= 0$ for $|x| \geq 1$. The support of ω_0 is the unit ball of R^n. By multiplying ω_0 by a suitable constant we obtain a nonnegative C^∞-function ω whose support is the unit ball of R^n, which is positive in the interior of this ball, and for which $\int \omega(x)\,dx = 1$. We will make repeated use of this function.

Lemma 1. If $u \in C_0^0(R^n)$ and if $\delta > 0$ is given, then we can find a function u_ε in $C_0^\infty(R^n)$ such that $|u(x) - u_\varepsilon(x)| \leq \delta$ for all x; i.e., $C_0^\infty(R^n)$ is dense in the Banach space $(C_0^0(R^n), \|\cdot\|_\infty)$.

Proof. Choose $\varepsilon > 0$ and define

$$u_\varepsilon(x) = \int u(x - \varepsilon y)\omega(y)\,dy = \varepsilon^{-n} \int u(y)\omega[(x - y)/\varepsilon]\,dy.$$

It is clear from the second of these that we are integrating over the compact set supp u. For $y \in$ supp u, $\omega[(x - y)/\varepsilon]$ is nonzero only in the ball of radius ε centered at y. Hence u_ε is in $C_0^0(R^n)$. But more is true. To compute any partial derivative of $u_\varepsilon(x)$ we differentiate $\omega[(x - y)/\varepsilon]$ under the integral sign. Hence, since ω is a C^∞-function, $u_\varepsilon(x) \in C_0^\infty(R^n)$.

Now

$$
\left| u_\varepsilon(x) - u(x) \right| \leq \varepsilon^{-n} \int |u(y) - u(x)| \, \omega\left(\frac{x - y}{\varepsilon}\right) dy
$$

$$
\leq \varepsilon^{-n}(\text{integral of } |u(y) - u(x)| \, \omega\left(\frac{x - y}{\varepsilon}\right)
$$

over $\{y \mid |u(y) - u(x)| \leq \delta\}) + \varepsilon^{-n}$ (integral of this same function over $\{y \mid |u(y) - u(x)| > \delta\}$). The first of these is

$$
\leq \delta \varepsilon^{-n} \int \omega[(x - y)/\varepsilon] \, dy \leq \delta.
$$

To estimate the second integral we note that, since it has compact support, u is uniformly continuous. Hence there is a $\gamma > 0$ such that $|u(y) - u(x)| > \delta$ implies $|x - y| > \gamma$. If we choose $\varepsilon > 0$ to be smaller than γ, then $\omega[(x - y)/\varepsilon] = 0$ for any y such that $|x - y| > \gamma$, i.e., for any y such that $|u(y) - u(x)| > \delta$. Thus for any such ε the second integral is zero.

We are going to prove that $C_0^\infty(R^n)$ is dense in $(L_p(R^n), \|\cdot\|_p)$ for any real $p \geq 1$. In order to do this we will need another lemma. First recall: A real-valued function ψ on R is said to be a convex function if for each $x_0 \in R$ there is an m in R such that $\psi(t) \geq \psi(x_0) + m(t - x_0)$ for all $t \in R$; i.e., the graph of ψ lies above the line that passes through $(x_0, \psi(x_0))$ and has slope m. The function $\psi(t) = |t|^p$ is a convex function if $p \geq 1$.

Lemma 2 (Jessen's Inequality). Let μ be a totally finite (i.e., $\mu(R^n) < \infty$), positive measure on R^n and let f be a real-valued, μ-integrable function on R^n. Then for any convex function ψ we have $\psi(\int f \, d\mu/\int d\mu) \leq \int \psi(f) \, d\mu/\int d\mu$.

Proof. Setting $x_0 = \int f \, d\mu/\int d\mu$ we can, since ψ is a convex function, find a number m such that $\psi(f) \geq \psi(x_0) + m(f - x_0)$. Hence $\int \psi(f) \, d\mu \geq \psi(x_0) \int d\mu + m \int f \, d\mu - mx_0 \int d\mu$ because μ is a positive measure.

Recalling how x_0 was defined we have

$$\int \psi(f)\, d\mu \geq \psi\left(\int f\, d\mu / \int d\mu\right) \int d\mu + m \int f\, d\mu - m\left(\int f\, d\mu / \int d\mu\right) \int d\mu.$$

Since μ is totally finite we can divide by $\int d\mu$.

Theorem 1.   For each real number $p \geq 1$ the space $C_0^\infty(R^n)$ is dense in the Banach space $(L_p(R^n), \|\cdot\|_p)$.

Proof.   Fix p and recall the nonnegative C^∞-function ω introduced just before Lemma 1. If $u \in L_p(R^n)$, then, for any positive ε, the function

$$u_\varepsilon(x) = \int u(x - \varepsilon y)\omega(y)\, dy = \varepsilon^{-n}\int u(y)\omega[(x-y)/\varepsilon]\, dy$$

is well defined because $\omega[(x - y)/\varepsilon]$ is in $L_q(R^n)$. We observe, as we did in the proof of Lemma 1, that $u_\varepsilon(x)$ is a C^∞-function with compact support in R^n. Thus, to prove the theorem, it suffices to show that $u_\varepsilon(x)$ tends to $u(x)$ for the L_p-norm as ε tends to zero.

The first thing that we want to establish is that $\|u_\varepsilon\|_p \leq \|u\|_p$. We have $\|u_\varepsilon\|_p^p = \int |\int u(x - \varepsilon y)\omega(y)\, dy|^p\, dx$. Set $\psi(t) = |t|^p$, recall that this is a convex function, and set $d\mu = \omega(y)\, dy$. Then, by Lemma 2,

$$\left|\frac{\int u(x - \varepsilon y)\omega(y)\, dy}{\int \omega(y)\, dy}\right|^p \leq \frac{\int |u(x - \varepsilon y)|^p \omega(y)\, dy}{\int \omega(y)\, dy}.$$

But since $\int \omega(y)\, dy = 1$, this says that $|\int u(x - \varepsilon y)\omega(y)\, dy|^p \leq \int |u(x - \varepsilon y)|^p \omega(y)\, dy$. Now we can write:

$$\|u_\varepsilon\|_p \leq \left[\iint |u(x - \varepsilon y)|^p \omega(y)\, dy\, dx\right]^{1/p}$$

$$= \left\{\int \omega(y)\left[\int |u(x - \varepsilon y)|^p\, dx\right] dy\right\}^{1/p}$$

$$= \left[\|u\|_p^p \int \omega(y)\, dy\right]^{1/p} = \|u\|_p.$$

Given $\eta > 0$ we choose a continuous function with compact support (call it v) such that $\|u - v\|_p < \eta$. Then clearly, $\|u_\varepsilon - v_\varepsilon\|_p < \eta$ and so

$$\|u - u_\varepsilon\|_p \leq \|u - v\|_p + \|v - v_\varepsilon\|_p + \|v_\varepsilon - u_\varepsilon\|_p < 2\eta + \|v - v_\varepsilon\|_p.$$

But, by Lemma 1, v_ε tends to v uniformly over R^n as ε tends to zero. Hence $\|v - v_\varepsilon\|_p$ tends to zero with ε and the theorem is proved.

We are now going to study C^∞-functions in Ω. Our main result, which will take a while to prove, is that there exists a C^∞-partition of unity subordinate to any open covering of Ω.

Lemma 3. Let $\{\Omega_v \,|\, v \in I\}$ be any open covering of Ω. Then there is a countable, open covering $\{u_k\}$ of Ω such that:

   (i)   Each u_k has compact closure.
   (ii) For each k there is a v with $u_k \subset \Omega_v$.
   (iii) Each point of Ω has a neighborhood that meets only a finite number of the sets $\{u_k\}$.

Proof. For each positive integer k let $K_k = \{x \in R^n \,|\, \text{distance of } x \text{ to } R^n \sim \Omega \text{ is } \geq 1/k \text{ and the distance from } x \text{ to zero is } \leq k\}$. Clearly each K_k is a compact set in Ω. Also, if int K_k denotes the interior of K_k, $K_k \subset \text{int } K_{k+1}$ for all k, and $\Omega = \bigcup_{k=1}^\infty K_k$.

Let $K_k = \varnothing$ for $k \leq 0$. For each fixed i, $L_i = K_i \cap \{\Omega \sim \text{int } K_{i-1}\}$ is a compact set and $V_i = \text{int } K_{i+1} \cap \{\Omega \sim K_{i-2}\}$ is an open neighborhood of this set. Each point x of L_i has an open neighborhood $W(x)$ that is contained in V_i and is also contained in some Ω_v. A finite number of these neighborhoods, say $W_{i,1}, W_{i,2}, \ldots, W_{i,l(i)}$, cover L_i. Choose such a finite open covering for each i. Since $W_{i,j} \subset K_{i+1}$ for each i and all j, each of these sets has compact closure. Thus $\{W_{i,j} \,|\, i = 1, 2, \ldots; 1 \leq j \leq l(i)\}$ is a countable open covering of Ω that has properties (i) and (ii).

We are going to show that $\{W_{i,j}\}$ has property (iii). Let $z \in \Omega$ and let i be the first integer for which $z \in \text{int } K_i$. Then $z \notin \text{int } K_{i-1}$ and so we can find a neighborhood V of z such that $V \subset \text{int } K_i$, $V \cap \text{int } K_{i-2} = \varnothing$. Let $y \in V \cap W_{m,j}$. Then $y \in W_{m,j}$ and so y is in int $K_{m+1} \cap \{\Omega \sim K_{m-2}\}$. Now if $m < i - 2$, then $m + 1 \leq i - 2$ and so y in int K_{i-2}. But $V \cap \text{int } K_{i-2} = \varnothing$. Thus m must be $\geq i - 2$. Also, if $i + 1 < m$, then $i - 1 < m - 2$, and so $i \leq m - 2$. Then $y \in \Omega \sim K_{m-2} \subset \Omega \sim K_i \subset \Omega \sim \text{int } K_i$. But $V \subset \text{int } K_i$, and so $m \leq i + 1$. We conclude that $V \cap W_{m,j} \neq \varnothing$ only for $i - 2 \leq m \leq i + 1$, and so V meets only a finite number of the sets $\{W_{i,j}\}$.

Definition 3. Let $\{\Omega_v \,|\, v \in I\}$ be any covering of Ω; we do not assume that the sets Ω_v are open, although they may be, and we do not assume that I is countable, although it may be. If every point of Ω has some neighborhood that meets only finitely many of the sets Ω_v, then we shall say that $\{\Omega_v\}$ is a locally finite covering of Ω.

Lemma 4. Let $\{U_k\}$ be a countable, locally finite, open covering of Ω. Then there is a countable, open covering $\{V_k\}$ of Ω such that the closure of V_k is contained in U_k for each k.

Proof. It suffices to find a sequence $\{V_k\}$ of open sets such that: (a) the closure of $V_k \subset U_k$ for every k; (b) for each positive integer m the sets V_k with $k \leq m$ and the sets U_k with $k > m$ cover Ω.

We shall define the sequence $\{V_k\}$ by induction. If U is any subset of Ω let cl U denote the closure of U in Ω. Suppose that we have defined the sets V_k with $k < l$ such that (a) is satisfied for $k < l$ and (b) is satisfied for $m < l$. Let $W = (\bigcup_{k<l} V_k) \cup (\bigcup_{k>l} U_k)$. Clearly, W is open, and note that $\Omega \sim U_l \subset W$ by (b) with $m = l - 1$. There is an open subset Z of Ω such that $\Omega \sim U_l \subset Z \subset \text{cl } Z \subset W$. If we set $V_l = \Omega \sim \text{cl } Z$, we have $V_l \subset \Omega \sim Z \subset U_l$. Also, since $\Omega \sim Z$ is closed in Ω, cl $V_l \subset U_l$. Finally, since $V_l \cup W = \Omega$, we see that (b) is satisfied for $m = l$.

One more lemma and we shall be able to prove our main theorem.

Lemma 5. Let $\{U_k\}$ be a countable, locally finite, open covering of Ω. Suppose that each U_k has compact closure. Then there is a family $\{\beta_k\} \subset C_0^\infty(\Omega)$ such that:

    (i) $\beta_k(x) \geq 0$ for all x in Ω and each $k = 1, 2, \ldots$.
    (ii) supp $\beta_k \subset U_k$ for each k.
    (iii) $\sum_{k=1}^\infty \beta_k(x) = 1$ for each $x \in \Omega$.

Proof. By Lemma 4 we can find an open covering $\{V_k\}$ of Ω such that cl $V_k \subset U_k$ for every k; here cl V_k denotes the closure of V_k. Fix k and, for each $x \in$ cl V_k, let $B(x)$ be a closed ball centered at x and contained in U_k. Since cl V_k is compact it can be covered by the interior of a finite number of these balls, say $B_1, B_2, \ldots, B_l$. For each i, $1 \leq i \leq l$, let x_i be the center and ε_i the radius of B_i. Let ω be the function introduced just before Lemma 1 and define $\rho_i(x)$ to be $\varepsilon_i^{-n}\omega[(x - x_i)/\varepsilon_i]$ for $i = 1, 2, \ldots, l$. Clearly, each ρ_i is a nonnegative function in $C_0^\infty(\Omega)$. Also, supp $\rho_i = B_i$ and $\rho_i > 0$ in the interior of B_i for each i. It follows that the function $\gamma_k(x) = \sum_{i=1}^l \rho_i(x)$ is in $C_0^\infty(\Omega)$, supp $\gamma_k \subset U_k$, and $\gamma_k > 0$ on cl V_k.

Now recall that $\{U_k\}$ is locally finite and that $\{V_k\}$ covers Ω. Thus $\gamma(x) = \sum_{k=1}^{\infty} \gamma_k(x)$ is a well-defined C^{∞}-function in Ω that is positive at each point of Ω. Clearly, the sequence $\beta_k(x) = \gamma_k(x)/\gamma(x)$, $k = 1, 2, \ldots$, satisfies conditions (i), (ii), and (iii).

Theorem 2. Let $\{\Omega_v \,|\, v \in I\}$ be an open covering of Ω. Then there is a family $\{\alpha_v \,|\, v \in I\}$ of C^{∞}-functions in Ω such that:

(a) $\alpha_v(x) \geq 0$ for all $x \in \Omega$ and each $v \in I$.
(b) $A_v \equiv \operatorname{supp} \alpha_v \subset \Omega_v$ for each $v \in I$.
(c) $\{A_v \,|\, v \in I\}$ is locally finite;
(d) $\sum \{\alpha_v(x) \,|\, v \in I\} = 1$ for each $x \in \Omega$.

The family $\{\alpha_v \,|\, v \in I\}$ is called a locally finite, C^{∞}-partition of unity subordinate to the covering $\{\Omega_v \,|\, v \in I\}$.

Proof. By Lemma 3 there is a countable, locally finite, open covering $\{U_k\}$ of Ω, with each U_k having compact closure, such that each U_k is contained in some Ω_v. Using $\{U_k\}$ and Lemma 5 we can find a sequence $\{\beta_k\} \subset C_0^{\infty}(\Omega)$ having properties (i), (ii), and (iii) of that lemma. For each positive integer k let $I(k) = \{v \in I \,|\, U_k \subset \Omega_v\}$. Each of these sets is nonempty and so, by the axiom of choice [15, Theorem 25, p. 33], there is a function T from the positive integers into I such that $T(k) \in I(k)$ for every k. For each $v \in I$ define $\alpha_v(x)$ to be $\sum \{\beta_k(x) \,|\, T(k) = v\}$. Clearly, $\{\alpha_v(x)\}$ is a family of C^{∞}-functions that has property (a).

To prove that $\{\alpha_v\}$ has property (b) first set $C_v = \bigcup \{\operatorname{supp} \beta_k \,|\, T(k) = v\}$. Clearly $C_v \subset \Omega_v$. Suppose that $x \in A_v \equiv \operatorname{supp} \alpha_v$. Then every neighborhood of x meets some set $\operatorname{supp} \beta_k$, where $T(k) = v$. On the other hand, some neighborhood of x meets only a finite number of the sets U_k, hence only a finite number of the sets $\operatorname{supp} \beta_k$. It follows that for some l, with $T(l) = v$, $x \in \operatorname{supp} \beta_l$. But then $x \in C_v$ and we have shown that $A_v \subset C_v \subset \Omega_v$ for each $v \in I$. Thus $\{\alpha_v\}$ has property (b). In order to prove that $\{\alpha_v\}$ has property (c) it suffices to prove that the family $\{C_v \,|\, v \in I\}$ is locally finite. For each $x \in \Omega$ there is a neighborhood V of x and a finite set H of positive integers such that $V \cap U_k = \varnothing$ for $k \notin H$. Then $V \cap \operatorname{supp} \beta_k = \varnothing$ for $k \notin H$. From this we see that if $v \notin \{T(k) \,|\, k \in H\}$ then, since C_v would be the union of sets $\operatorname{supp} \beta_k$, where $k \notin H$, $V \cap C_v = \varnothing$. But since $\{T(k) \,|\, k \in H\}$ is a finite set we have shown that $\{C_v\}$ is a locally finite family.

Finally, for each $x \in \Omega$,

$$1 = \sum_{k=1}^{\infty} \beta_k(x) = \sum_{v \in I} \left[\sum_{T(k)=v} \beta_k(x) \right] = \sum_{v \in I} \alpha_v(x).$$

Corollary 1.   Let O, C be two subsets of Ω, O an open set, C a closed set, and suppose that $C \subset O \subset \Omega$. Then there is a function $\phi \in C^{\infty}(\Omega)$ such that:

(i) $0 \le \phi(x) \le 1$ for all $x \in \Omega$;
(ii) $\phi(x) = 1$ for all $x \in C$.
(iii) $\phi(x) = 0$ for all $x \in \Omega \sim O$.

Proof.   There is a C^{∞}-partition of unity (call it α_1, α_2) that is subordinate to the open covering $O \equiv \Omega_1$, $\Omega \sim C \equiv \Omega_2$. Clearly $\phi(x) = \alpha_1(x)$ for all x has properties (i), (ii), and (iii).

EXERCISES 1

*1.   A multi-index is an ordered n-tuple of nonnegative integers. If $s = (s_1, s_2, \ldots, s_n)$ is a multi-index we let $|s| = s_1 + s_2 + \cdots + s_n$ and, for each f in $C^{|s|}(R^n)$, we let

$$D^s f = \frac{\partial^{|s|} f}{\partial x_1^{s_1} \, \partial x_2^{s_2} \cdots \partial x_n^{s_n}}.$$

Let K be a fixed, compact subset of R^n and let $\mathscr{D}_K(R^n) = \{f \in C_0^{\infty}(R^n) \,|\, \mathrm{supp}\, f \subset K\}$. For each nonnegative integer m define p_m as follows: $p_m(f) = \sup\{|D^s f(x)| \,|\, x \in K,\ |s| \le m\}$ for each $f \in \mathscr{D}_K(R^n)$.

(a)   Show that each p_m is a seminorm on $\mathscr{D}_K(R^n)$ (Section 4.2, Definition 3).

(b)   If $f \in \mathscr{D}_K(R^n)$ is not the zero function show that $p_m(f) \ne 0$ for some m.

(c)   The family $\{p_m \,|\, m = 0, 1, 2, \ldots\}$ defines a Hausdorff, locally convex topology on $\mathscr{D}_K(R^n)$ (Section 4.1; see the construction process just after Lemma 1). Call this topology $t(\{p_m\})$. Show that there is a countable fundamental system of $t(\{p_m\})$-neighborhoods of zero in $\mathscr{D}_K(R^n)$.

*2. Let X be a vector space over K and let $\{p_n\}$ be a countable family of seminorms on X. For each positive integer m define $q_m(x)$ to be $\sup\{p_j(x) \mid 1 \le j \le m\}$ for each $x \in X$.

   (a) Show that each q_m is a seminorm on X and that $q_m(x) \le q_{m+1}(x)$ for all $x \in X$ and each $m = 1, 2, \dots$.

   (b) If $\{p_n\}$ satisfies the separation condition (Section 4.1, Lemma 2) then so does $\{q_n\}$, and conversely.

   (c) Show that the families $\{p_n\}$ and $\{q_n\}$ define equivalent topologies on X. Hint: It suffices to show that every $t(\{p_n\})$-neighborhood of zero in X contains a $t(\{q_n\})$-neighborhood of zero, and conversely.

3. (a) Let Ω be any open subset of R^n, and let C be a closed subset of Ω and O an open subset of Ω such that $C \subset O$. Show that there is an open subset W of Ω such that $C \subset W \subset \operatorname{cl} W \subset O$.

   (b) Prove that the union of any locally finite family of closed subsets of R^n is a closed subset of R^n.

2. Fréchet Spaces

Locally convex spaces whose topologies can be defined by means of a countable family of seminorms arise frequently enough to warrant some special attention. The one example we have seen, $\mathscr{D}_K(R^n)$ (defined in Exercises 1, problem 1) will be useful later on. Another useful example will be given at the end of this section.

Theorem 1. Let X be a vector space over K and let t be a Hausdorff, locally convex topology on X. Then the following are equivalent:

(a) The topology t is metrizable.

(b) There is in X a countable, fundamental system of t-neighborhoods of zero.

(c) The topology t can be defined by means of a countable family of seminorms that satisfies the separation condition.

Proof. It is clear that (a) implies (b). Assume (b) and let $\{U_n\}$ be a countable, fundamental system of t-neighborhoods of zero in X such that $U_n \supset U_{n+1}$ for all n. We may assume that each U_n is absorbing, balanced, and convex (Section 4.3, Lemma 2). Let p_n be the gauge function of U_n for each n (Section 4.2, just before Definition 1). Then (Section 4.2, Theorem 1) each p_n is a seminorm on X. Clearly, the topology defined by the family $\{p_n\}$ coincides with t (Section 4.3, proof of Theorem 1). To prove that (c) implies (a) we need:

Lemma 1.   Let s be a Hausdorff, locally convex topology on X defined by a sequence of seminorms $\{p_n\}$ such that $p_n(x) \leq p_{n+1}(x)$ for all x and all n. For each $x \in X$ define $|x|$ to be $\sum_{n=1}^{\infty} 2^{-n} p_n(x) \times (1 + p_n(x))^{-1}$. Then:

(i) $|x| = 0$ iff $x = 0$.
(ii) $|x| = |-x|$ for each $x \in X$.
(iii) $|x + y| \leq |x| + |y|$ for all x, y in X.
(iv) For any $x \in X$ and any $\lambda \in K$ such that $|\lambda| \leq 1$, $|\lambda x| \leq |x|$.
(v) If the sequence $\{\lambda_n\}$ (of scalars) converges to zero, then, for each $x \in X$, $\{|\lambda_n x|\}$ converges to zero.

Finally, if we define $\rho(x, y)$ to be $|x - y|$ for all $x, y \in X$, then ρ is a metric on X and ρ is translation invariant; i.e., $\rho(x + z, y + z) = \rho(x, y)$ for all x, y, z in X.

Proof.   Properties (i) and (ii) of $|\cdot|$ and the fact that $|x| \geq 0$ for all x, are obvious. The function that takes each real number α to $\alpha(1 + \alpha)^{-1}$ is an increasing function for $\alpha \neq -1$ (just take its derivative). Hence:

(1) $$\alpha(1 + \alpha)^{-1} \leq \beta(1 + \beta)^{-1} \qquad \text{for} \quad -1 < \alpha \leq \beta.$$

If $\lambda \in K$, $|\lambda| \leq 1$, then $p_n(\lambda x) = |\lambda| p_n(x) \leq p_n(x)$ (Section 4.1, Definition 1). Using (1) we may write

$$p_n(\lambda x)(1 + p_n(\lambda x))^{-1} \leq p_n(x)(1 + p_n(x))^{-1}$$

and property (iv) follows from this.

We shall use (1) to prove (iii). Recall that $p_n(x + y) \leq p_n(x) + p_n(y)$ (Section 4.1, Definition 1). Hence

$$p_n(x + y)(1 + p_n(x + y))^{-1} \leq (p_n(x) + p_n(y))(1 + p_n(x) + p_n(y))^{-1}$$
$$\leq p_n(x)(1 + p_n(x))^{-1} + p_n(y)(1 + p_n(y))^{-1}$$

and (iii) follows from this.

Properties (i), (ii), and (iii) imply that $\rho(x, y) = |x - y|$ is a translation invariant metric on X. Let s' be the topology defined on X by ρ. We shall investigate the relationships between the s- and the s'-neighborhoods of zero in X.

Let $U = \{x \in X \mid |x| \leq 2^{-k}\}$. We shall show that U contains the set $V = \{x \in X \mid p_{k+1}(x) \leq 2^{-(k+1)}\}$. If $x \in V$ then $p_1(x) \leq p_2(x) \leq \cdots \leq p_{k+1}(x) \leq 2^{-(k+1)}$ and so, since $p_n(x)(1 + p_n(x))^{-1} \leq p_n(x)$,

$$(2) \qquad \sum_{n=1}^{k+1} 2^{-n} p_n(x)(1 + p_n(x))^{-1} \leq 2^{-(k+2)} \sum_{n=1}^{\infty} 2^{-n} = 2^{-(k+3)}.$$

On the other hand $p_n(x)(1 + p_n(x))^{-1} \leq 1$ and so

$$(3) \qquad \sum_{k+2}^{\infty} 2^{-n} p_n(x)(1 + p_n(x))^{-1} \leq \sum_{k+2}^{\infty} 2^{-n} = 2^{-(k+2)}.$$

Combining (2) and (3) we get $|x| \leq 2^{-k}$ and so $x \in U$. Thus s is stronger than s'.

Now let $W = \{x \in X \mid p_m(x) \leq 2^{-(m+k+1)}\}$. We shall show that W contains $W' = \{x \in X \mid |x| \leq 2^{-(m+k+1)}\}$. If $x \in W'$, then certainly $2^{-m} p_m(x)(1 + p_m(x))^{-1} \leq 2^{-(m+k+1)}$. From this it follows that $p_m(x) \times (1 + p_m(x))^{-1} \leq 2^{-(k+1)}$ and that $p_m(x)(1 - 2^{-(k+1)}) \leq 2^{-(k+1)}$. Thus $p_m(x) \leq 2^{-(k+1)} - 1 \leq 2^{-k}$, which says $x \in W$.

If $\{\lambda_n\}$ is a sequence of scalars that converges to zero, then for each $x \in X$ the sequence $\{\lambda_n x\}$ is s-convergent to zero. But then $\{\lambda_n x\}$ is s'-convergent to zero and this proves (v).

Since every s-neighborhood of zero contains an s'-neighborhood of zero, and conversely, and since ρ is translation invariant, the topologies s and s' must coincide on X.

We now return to the proof of Theorem 1. Assume (c) and let $\{p_n\}$ be a sequence of seminorms such that $t = t(\{p_n\})$. By problem 2 of Exercises 1, there is an increasing sequence of seminorms $\{q_n\}$ on X such that $t = t(\{q_n\})$. But, by Lemma 1, $t(\{q_n\})$ is metrizable.

Definition 1. A locally convex space whose topology is metrizable will be called a metrizable, locally convex space. A complete, metrizable, locally convex space (i.e., one in which every sequence that is Cauchy for the metric converges for the metric to a point of the space) will be called a Fréchet space.

Remark. Let $X[t]$ be a locally convex space. We recall (Section

4.3, Definition 3) that a subset B of X is said to be t-bounded if for each t-neighborhood, say V, of zero in X there is a scalar λ such that $B \subset \lambda U$. Suppose that t is a metrizable topology and that ρ is a metric that defines t. A set $B \subset X$ is bounded for the metric if $\sup\{\rho(x, y) \mid x, y$ in $B\}$ is finite. Note that such a set need not be t-bounded. In fact, Lemma 1 shows that X can be bounded for the metric. Whenever we speak of a bounded subset of a metrizable, locally convex space $X[t]$ we shall always mean a set that is t-bounded in X.

Let $s = (s_1, s_2, \ldots, s_n)$ be a multi-index (Exercises 1, problem 1). We define $s!$ to be $s_1! \, s_2! \ldots s_n!$ and, if $x = (x_1, \ldots, x_n)$, x^s to be $x_1^{s_1} x_2^{s_2} \cdots x_n^{s_n}$. If m is any positive integer then $(x_1 + x_2 + \cdots + x_n)^m = \sum \{(m!/s!)x^s \mid |s| = m\}$.

Lemma 2.   For a C^∞-function f on R^n the two following conditions are equivalent:

(1)   For any positive integer k, any multi-index s, and any $\varepsilon > 0$, there is a number $\sigma > 0$ such that $\left|(1 + |x|^2)^k D^s f(x)\right| \leq \varepsilon$ for $|x| > \sigma$.

(2)   For any two multi-indices r and s and any $\varepsilon > 0$ there is a number $\sigma > 0$ such that $\left|x^r D^s f(x)\right| \leq \varepsilon$ for $|x| > \sigma$.

Proof.   The expression $(1 + |x|^2)^k$ is a linear combination of monomials of the form $x^{2r} = x_1^{2r_1} x_2^{2r_2} \cdots x_n^{2r_n}$, with $|r| \leq k$. Hence, for any multi-index s, there is a constant M such that

$$\max\{\left|(1 + |x|^2)^k D^s f(x)\right| \mid x \in R^n\}$$
$$\leq M \max \max\{\left|x^r D^s f(x)\right| \mid |r| \leq 2k, x \in R^n\}.$$

Thus (2) implies (1). But since $|x^r| \leq (1 + |x|^2)^k$ for $|r| \leq 2k$, we see that (1) implies (2).

Definition 2.   A C^∞-function on R^n is said to be a rapidly decreasing function on R^n if it satisfies condition (2) of Lemma 2. The set of all rapidly decreasing functions on R^n will be denoted by $\mathscr{S}(R^n)$.

It is obvious that $\mathscr{S}(R^n)$ is a space of functions on R^n (Section 1.2, paragraph before Exercises 2) and that $C_0^\infty(R^n) \subset \mathscr{S}(R^n)$. Also, if f is an element of $\mathscr{S}(R^n)$, then so is any partial derivative, of any order, of f.

We can define two countable families of seminorms on $\mathscr{S}(R^n)$. First, for any two multi-indices r and s, we can define $q_{r,s}(f)$ to be

$\max\{|x^r D^s f(x)| \,|\, x \in R^n\}$ for all $f \in \mathcal{S}(R^n)$. It is obvious that each $q_{r,s}$ is a seminorm on $\mathcal{S}(R^n)$, that $\{q_{r,s}\}$ is a countable family, and that this family satisfies the separation condition. The second family is defined as follows: For each positive integer k and each multi-index s let $p_{k,s}(f) = \max\{|(1 + |x|^2)^k D^s f(x)| \,|\, x \in R^n\}$ for all $f \in \mathcal{S}(R^n)$. The family $\{p_{k,s}\}$ is a countable family of seminorms that satisfies the separation condition. By Theorem 1 each of these families defines a metrizable, locally convex topology on $\mathcal{S}(R^n)$. The proof of Lemma 2 shows that these two topologies are equivalent.

From here on we shall always assume, often without explicit mention, that $\mathcal{S}(R^n)$ has the metrizable, locally convex topology defined by one or the other of the families of seminorms discussed above. We shall denote this topology by t_d.

Theorem 2. The space $\mathcal{S}(R^n)$ is a Fréchet space, and $C_0^\infty(R^n)$ is dense in this space.

Proof. Let $\{f_m\}$ be a t_d-Cauchy sequence of points of $\mathcal{S}(R^n)$. Then $\lim q_{r,s}(f_n - f_m) = 0$, as m, n tend to infinity, for any r and s. Let $|r| = 0$. Then, by letting $|s| = 0, 1, 2, \ldots$, we see that $\{f_m\}$, and each $\{D^s f_m\}$, is uniformly Cauchy over R^n. So $\{f_m\}$ converges, uniformly over R^n, to a function f. The function $f \in C^\infty(R^n)$ and, for every s, $\{D^s f_m\}$ converges uniformly over R^n to $D^s f$. It follows that $\{x^r D^s (f - f_m)\}$ tends to zero, uniformly over R^n, as m tends to infinity, for every fixed pair r, s. Hence the inequality

$$|x^r D^s f(x)| \leq |x^r D^s (f - f_m)(x)| + |x^r D^s f_m(x)|$$

shows that $f \in \mathcal{S}(R^n)$.

In order to prove that $C_0^\infty(R^n)$ is dense in $\mathcal{S}(R^n)$ we first use the Corollary to Theorem 2 (Section 1) to find a C^∞-function γ such that $0 \leq \gamma(x) \leq 1$ for all x, $\gamma(x) = 1$ if $|x| \leq 1$, and $\gamma(x) = 0$ if $|x| > 2$. Let $f \in \mathcal{S}(R^n)$ and $\varepsilon > 0$ be given. Let k be a positive integer, let s be a multi-index, and for any integer j let $\gamma_j(x) = \gamma(x/j)$. Then

$$|(1 + |x|^2)^k D^s \{ f(x)(1 - \gamma_j(x)) \}|$$
$$\leq \sum \binom{s}{r} |D^{s-r}(1 - \gamma_j(x))| \, |(1 + |x|^2)^k D^r f(x)|,$$

where $\binom{s}{r} = s!/r!(s-r)!$ and $s - r = (s_1 - r_1, s_2 - r_2, \ldots, s_n - r_n)$. Suppose that we have l summands. There is a constant C such that $|D^{s-r}(1 - \gamma_j(x))| \leq C$ for all x; C is independent of j. Also, there is a

$\sigma > 0$ such that $\left| (1 + |x|^2)^k D^r f(x) \right| \le \varepsilon/(lC)$ for $|x| > \sigma$ and $|r| \le |s|$. By choosing $j > \sigma$ we have $\left| (1 + |x|^2)^k D^s(f(x) - f(x)\gamma_j(x)) \right| \le \varepsilon$ for all x.

EXERCISES 2

1. Show that $\mathscr{D}_K(R^n)$ in problem 1 of Exercises 1, given the topology defined by the countable family of seminorms also discussed in that problem, is a Fréchet space.

2. Let $X[t]$ be a metrizable, locally convex space and let $\{B_n\}$ be any sequence of t-bounded subsets of X. Show that there is a sequence of positive numbers $\{\lambda_n\}$ such that $\bigcup_{n=1}^{\infty} \lambda_n B_n$ is a t-bounded set. Hint: First show that we can assume that $B_n \subset B_{n+1}$ for all n. Next let $\{U_n\}$ be a decreasing, fundamental sequence of t-neighborhoods of zero in X and, for each n, choose λ_n so that $\lambda_n B_n \subset U_n$.

*3. (a) Show that $\exp(-|x|^2/2)$ is in $\mathscr{S}(R)$. Hint: Use L'Hospital's rule.

   (b) If $\phi \in \mathscr{S}(R)$ show that $\left| \int \phi(x) \exp(-ixy) \, dx \right|$ is finite for every fixed y.

   (c) Prove that $\int e^{-x^2/2} \, dx = (2\pi)^{1/2}$ as follows: First note that $\left[\int e^{-x^2/2} \, dx \right]^2 = \iint \exp[-(x^2 + y^2)/2] \, dx \, dy$. Then evaluate this double integral using polar coordinates.

3. **The Fourier Transform**

For f in $L_1[-\pi, \pi]$ we defined the Fourier transform $\hat{f}$ of f to be $(1/2\pi) \int f(x) e^{-inx} \, dx$, $n = 0, \pm 1, \ldots$ (Section 6.1, Definition 1). We saw that $\hat{f}$ has some useful properties and we found that the map that takes each $f \in L_1[-\pi, \pi]$ to $\hat{f}$ has an inverse (Section 6.1, Theorem 3). A similar "transform," also called the Fourier transform, can be defined for many classes of functions, and even more general objects. For example, if $f \in L_1(R)$, then we may define $\hat{f}(y)$ to be $(1/2\pi)^{1/2} \int f(x) e^{-ixy} \, dx$. (Some writers leave out the factor $(1/2\pi)^{1/2}$, and others replace it by $1/2\pi$. For a discussion of this point see [14, p. 122].) Clearly, $|\hat{f}(y)| \le (1/2\pi)^{1/2} \|f\|_1$ and, by trivially modifying the proof of the Riemann–Lebesgue theorem (Section 6.1, Theorem 1), one can show that $\lim_{y \to \pm\infty} \hat{f}(y) = 0$. Also, since $|\hat{f}(y + z) - \hat{f}(y)| \le \int |f(x)| |e^{ixz} - 1| \, dx$ and the integrand is bounded by $2|f(x)|$ and tends to zero everywhere as $z \to 0$, $\hat{f}$ is uniformly continuous. So we have a map, $f \to \hat{f}$, that takes each $f \in L_1(R)$ to a function that is uniformly continuous and vanishes at infinity on R. Note that it is helpful to have two real lines in mind here. One of them, R, is the domain of our L_1-functions and the other one (call it $\hat{R}$) is the domain of their Fourier transforms.

Now what properties does the map $f \to \hat{f}$ have? In particular, does it have an inverse? Also, since R has infinite measure, $L_p(R)$ is not contained in $L_1(R)$ when $p > 1$. So how would we define the Fourier transform for $f \in L_p(R)$, $p > 1$? We shall be able to answer these questions at the end of this section. A more detailed treatment of the Fourier transform on $L_1(R)$ can be found in [8] or [14]. The latter also contains a discussion of the Fourier transform on $L_p(R)$. For a much more general treatment of the entire subject see [9].

Definition 1. For each function ϕ in $\mathscr{S}(R)$ let $\hat{\phi}(y) = (1/2\pi)^{1/2} \int \phi(x)e^{-ixy} \, dx$. We regard the function $\hat{\phi}(y)$ (Exercises 2, problem 3b) as defined on a second copy $\hat{R}$ of R.

Lemma 1. For any function $\phi \in \mathscr{S}(R)$ the function $\hat{\phi} \in \mathscr{S}(\hat{R})$. The map, call it $\mathscr{F}$, that takes each function $\phi \in \mathscr{S}(R)$ to the function $\hat{\phi} \in \mathscr{S}(\hat{R})$ is linear and continuous.

Proof. Choose $\phi \in \mathscr{S}(R)$. We shall first show that $\hat{\phi}$ is a C^∞-function. For any p we have:

$$(*) \qquad D^p\hat{\phi}(y) = \frac{(-i)^p}{(2\pi)^{1/2}} \int x^p\phi(x)e^{-ixy} \, dx.$$

We may write

$$|x^p\phi(x)| = \left| \frac{x^p(1 + x^2)^{p+k}\phi(x)}{(1 + x^2)^{p+k}} \right|,$$

where k is an integer, and if $|x| \ge 1$, then

$$(**) \qquad |x^p\phi(x)| \le \frac{|x|^p}{(1 + x^2)^{p+k}} \max |(1 + x^2)^{p+k}\phi(x)|.$$

However, $\phi \in \mathscr{S}(R)$, and so the maximum is finite. Thus, since $\int (|x|^p/(1 + x^2)^{p+k})\, dx$ is clearly finite, $\hat{\phi}$ is a C^∞-function.

Return now to the definition of $\hat{\phi}$ and integrate by parts. We get

$$\hat{\phi}(y) = \frac{1}{(2\pi)^{1/2}} \left(\frac{\phi(x)e^{-ixy}}{-iy} \right) \Bigg|_{-\infty}^{\infty} - \frac{1}{(2\pi)^{1/2}} \int \frac{\phi'(x)e^{-ixy}}{-iy}\, dx.$$

The first term on the right is zero because $\phi \in \mathscr{S}(R)$. Repeated integration by parts gives $(iy)^q \hat{\phi}(y) = (1/2\pi)^{1/2} \int D^q \phi(x) e^{-ixy}\, dx$. Combining this equation with equation $(*)$ above we get

$$(***) \qquad (iy)^q D^p \hat{\phi}(y) = \frac{(-i)^p}{(2\pi)^{1/2}} \int D^q[x^p \phi(x)] e^{-ixy}\, dx.$$

Hence, for any $\sigma > 0$, $|y^q D^p \hat{\phi}(y)| \leq |\int D^q[x^p \phi(x)]e^{-ixy}\, dx|$ over $\{x \mid |x| \leq \sigma\}$ plus $\int |D^q[x^p \phi(x)]|\, dx$ over $\{x \mid |x| > \sigma\}$. We shall use this inequality to show that, for any $\varepsilon > 0$, $|y^q D^p \hat{\phi}(y)|$ is less than ε for all y sufficiently large. This will prove that $\hat{\phi} \in \mathscr{S}(\hat{R})$ (Section 2, Definition 2). The first term on the right-hand side of our inequality is of the form $\int_{-\sigma}^{\sigma} f(x)e^{-ixy}\, dx$, where σ is fixed and f is differentiable. The proof of the Riemann–Lebesgue theorem (Section 6.1, Theorem 1) shows that this term tends to zero as $y \to \pm\infty$. We will now show that the second term can be made as small as we please by taking σ sufficiently large. At the same time we shall see that $\mathscr{F}$ is continuous.

Suppose first that $p = 3$, $q = 2$. Then $\int |D^q[x^p \phi(x)]|\, dx$ over $\{x \mid |x| > \sigma\}$ becomes

$$\int |D^2[x^3 \phi(x)]|\, dx$$

$$\leq \int |x^3 \phi''(x)|\, dx + 6 \int |x^2 \phi'(x)|\, dx + 6 \int |x\phi(x)|\, dx$$

(by the Leibnitz formula). Using inequality $(**)$ we find that

$$\int |D^2[x^3 \phi(x)]|\, dx$$

$$\leq \max\{|(1 + x^2)^{3+k} \phi''(x)| \mid |x| \geq \sigma\} \int \frac{|x|^3}{(1 + x^2)^{3+k}}\, dx$$

plus two similar terms. Since each of the functions ϕ, ϕ', ϕ'' is in $\mathscr{S}(R)$ and, for k sufficiently large, each of the integrals is convergent we see that $\int |D^2[x^3 \phi(x)]|\, dx$ can be made as small as we please by taking σ

sufficiently large. Also, recalling the definitions of the seminorms $q_{r,s}$ and $p_{k,s}$ (Section 2, just after Definition 2), we see that: $q_{2,3}(\hat{\phi}) = \max|y^3 D^2 \hat{\phi}(y)| \leq M p_{3+k,2}(\phi)$ plus two similar terms; here M is a constant that is greater than $\int (|x|^3/(1 + x^2)^{3+k})\,dx$. It follows that $\mathscr{F}$ is continuous.

The proof for the general case is similar to the one just given because, by the Leibnitz formula, $D^q[x^p\phi(x)]$ is a linear combination of terms $x^r D^s \phi(x)$.

It will be useful to know the Fourier transform of $e^{-x^2/2}$. This function is in $\mathscr{S}(R)$ (Exercises 2, problem 3a) and we shall need the formula proved in part (c) of that problem.

Lemma 2.

$$(1/2\pi)^{1/2} \int e^{-x^2/2} e^{-ixy}\,dx = e^{-y^2/2}.$$

Proof. First write

(1) $$\int_{-\lambda}^{\lambda} e^{-t^2/2} e^{-iut}\,dt = e^{-u^2/2} \int_{-\lambda}^{\lambda} \exp\left[\frac{-(t+iu)^2}{2}\right] dt.$$

For any $v > 0$ the Cauchy integral theorem shows that $\int_C \exp(-z^2/2)\,dz = 0$ if C is the rectilinear path joining $-\lambda$ to λ, λ to $\lambda + iv$, $\lambda + iv$ to $-\lambda + iv$, and $-\lambda + iv$ to $-\lambda$. Hence we may write

$$0 = \int_{-\lambda}^{\lambda} \exp(-t^2/2)\,dt + \int_0^v \exp[-(\lambda + iu)/2]i\,dv$$

$$+ \int_{\lambda}^{-\lambda} \exp[-(t + iv)^2/2]\,dt + \int_v^0 \exp[-(-\lambda + iu)^2/2]i\,du.$$

Notice that the second and fourth terms in this equation tend to zero as λ tends to infinity. Transposing the third term we find that

$$\int_{-\infty}^{\infty} \exp[-(t + iv)^2/2]\,dt = \int_{-\infty}^{\infty} \exp(-t^2/2)\,dt.$$

Combining this with (1),

$$\int_{-\infty}^{\infty} e^{-t^2/2} e^{-iut}\,dt = e^{-u^2/2} \int_{-\infty}^{\infty} e^{-t^2/2}\,dt = (2\pi)^{1/2} e^{-u^2/2}.$$

Finally, if $\phi(x) = e^{-x^2/2}$, then

$$\hat{\phi}(y) = \frac{1}{\sqrt{2\pi}} \int \phi(x)e^{-ixy}\, dx = \frac{\sqrt{2\pi}}{\sqrt{2\pi}} e^{-y^2/2} = e^{-y^2/2}.$$

We want to prove that the map $\mathscr{F}$, defined in the statement of Lemma 1, has a continuous inverse. To do that we shall need Lemma 2 and the next result.

Theorem 1 (Parseval Relations).   For any two functions ϕ, ψ in $\mathscr{S}(R)$ and $\mathscr{S}(\hat{R})$, respectively, we have $\int \hat{\phi}(y)\psi(y)\, dy = \int \phi(x)\hat{\psi}(x)\, dx$.

Proof.

$$\int \hat{\phi}(y)\psi(y)e^{ixy}\, dy = \int \psi(y)\left[(1/2\pi)^{1/2} \int \phi(z)e^{-iyz}\, dz\right]e^{ixy}\, dy$$

$$= \int \left[(1/2\pi)^{1/2} \int \psi(y)e^{-iy(z-x)}\, dy\right]\phi(z)\, dz$$

$$= \int \hat{\psi}(z-x)\phi(z)\, dz,$$

where we now regard ψ as a function in $\mathscr{S}(R)$. Setting $w = z - x$ in this last expression we find:

(∗) $$\int \hat{\phi}(y)\psi(y)e^{ixy}\, dy = \int \hat{\psi}(w)\phi(w+x)\, dw.$$

Our result follows from (∗) upon setting $x = 0$.

Observe that, if $\psi \in \mathscr{S}(R)$, for any $\varepsilon > 0$ the function $\phi(x) = \psi(\varepsilon x)$ is in $\mathscr{S}(R)$ and $\hat{\phi}(y) = (1/\varepsilon)\hat{\psi}(y/\varepsilon)$.

Theorem 2.   The map $\mathscr{F}$ that takes each element of $\mathscr{S}(R)$ to its Fourier transform has an inverse that is linear and continuous. In particular, $\mathscr{F}$ is one-to-one and onto.

Proof.   Define a linear map from $\mathscr{S}(\hat{R})$ to $\mathscr{S}(R)$ as follows: For each ψ in $\mathscr{S}(\hat{R})$ let $\tilde{\psi}(x) = (1/2\pi)^{1/2} \int \psi(y)e^{ixy}\, dy$, and let our map take ψ to $\tilde{\psi}$. Anticipating our result, we shall denote this map by $\mathscr{F}^{-1}$. It is clear that $\mathscr{F}^{-1}$ is linear and the proof of Lemma 1 shows that it is continuous. So all we have to do is prove that $\mathscr{F}$ and $\mathscr{F}^{-1}$ really are inverses of each other.

For ψ in $\mathscr{S}(\hat{R})$ and any fixed $\varepsilon > 0$ let $\alpha(y) = \psi(\varepsilon y)$. Then, as we observed above, $\hat{\alpha}(z) = \varepsilon^{-1}\hat{\psi}(z/\varepsilon)$. Equation $(*)$ in the proof of Theorem 1 reads, in this case, $\int \alpha(y)\hat{\phi}(y)e^{ixy}\,dy = \int \hat{\alpha}(w)\phi(w + x)\,dw$ and this is equal to $\varepsilon^{-1} \int \hat{\psi}(\varepsilon^{-1}w)\phi(w + x)\,dw = \int \hat{\psi}(u)\phi(x + \varepsilon u)\,du$ where $u = \varepsilon^{-1}w$. If, in particular, we take $\psi(x)$ to be $\exp(-x^2/2)$, then

$$\int \exp[-(\varepsilon y)^2/2]\hat{\phi}(y)e^{ixy}\,dy = \int \hat{\psi}(u)\phi(x + \varepsilon u)\,du.$$

In this equation let $\varepsilon \to 0$. We find:

$$\int \hat{\phi}(y)e^{ixy}\,dy = \phi(x) \int \hat{\psi}(u)\,du = \phi(x) \int \exp(-u^2/2)\,du = (2\pi)^{1/2}\phi(x),$$

by Lemma 2. Thus $(2\pi)^{1/2}[\hat{\hat{\phi}}(x)] = (2\pi)^{1/2}\phi(x)$, which proves the theorem.

Theorem 3 (also called Parseval's Relation). For any two functions ϕ, ψ in $\mathscr{S}(R)$ we have $\int \phi(y)\bar{\psi}(y)\,dy = \int \hat{\phi}(x)\overline{\hat{\psi}}(x)\,dx$, where the bar denotes complex conjugation.

Proof. By Theorem 1, $\int \hat{\phi}(y)\psi(y)\,dy = \int \phi(x)\hat{\psi}(x)\,dx$. Since $\phi(y) = \mathscr{F}[\hat{\phi}(x)]$ we need only show that $\bar{\psi}(x) = \mathscr{F}[\overline{\hat{\psi}}(y)]$. But $\mathscr{F}[\overline{\hat{\psi}}(y)] = (1/2\pi)^{1/2} \int \overline{\hat{\psi}}(x)e^{-ixy}\,dx = $ complex conjugate of

$$(1/2\pi)^{1/2} \int \psi(x)e^{ixy}\,dy = \bar{\psi}(x).$$

At this point the reader may well wonder what it is that we are doing. After all, it is clear that $L_p(R)$ is not contained in $\mathscr{S}(R)$. So how is the Fourier transform on $\mathscr{S}(R)$ going to help us answer the questions raised earlier? The answer is rather interesting. What we shall do is show that $L_p(R)$ is contained in the "dual" of the locally convex space $\mathscr{S}(R)$. Our results so far, about $\mathscr{S}(R)$, will enable us to define a Fourier transform on this dual. Thus, in particular, we get a Fourier transform on $L_p(R)$ this way. We shall make all this precise now.

Remark. Let $X[t]$ be a locally convex space. The vector space of all t-continuous, linear functionals on X will be called the dual space of $X[t]$ and will be denoted by X'. X' is not the trivial vector space (Section 4.4, Corollary 1 to Theorem 1). For each $f \in X'$ let $p_f(x) = |f(x)|$ for all $x \in X$. The family $\{p_f \mid f \in X'\}$ defines a locally convex topology on X (just as in Section 4.1), which we shall call the weak topology on X and denote by $\sigma(X, X')$. Similarly, we define a weak* topology $\sigma(X', X)$ on X' by means of the family $\{p_x \mid x \in X\}$, where $p_x(f) = |f(x)|$ for all $f \in X'$.

Definition 2. A continuous, linear functional on $\mathscr{S}(R)$ will be called a temperate distribution in R. The set of all temperate distributions in R (i.e., the dual of $\mathscr{S}(R)$) will be denoted by $\mathscr{S}'(R)$.

Lemma 3. For $T \in \mathscr{S}'(\hat{R})$ let $\hat{T}(\phi) = T(\hat{\phi})$ for every $\phi \in \mathscr{S}(R)$. Then $\hat{T}$ is in $\mathscr{S}'(R)$. Furthermore, the map $\mathscr{G}$ that takes each $T \in \mathscr{S}'(\hat{R})$ to $\hat{T} \in \mathscr{S}'(R)$ is continuous when these two spaces have their weak* topologies.

Proof. If $T \in \mathscr{S}'(\hat{R})$ then $\hat{T}(\phi) = T(\hat{\phi}) = T \circ \mathscr{F}(\phi)$. Since both $\mathscr{F}$ and T are continuous, $\hat{T}$ is continuous, i.e., $\hat{T} \in \mathscr{S}'(R)$.

Since $\mathscr{G}$ is obviously linear we shall have proved that it is continuous once we have shown that it is continuous at zero. Let U be an arbitrary weak* neighborhood of zero in $\mathscr{S}'(R)$. We may assume that $U = \{T \in \mathscr{S}'(R) \mid |T(\phi_j)| \le \varepsilon_j \text{ for } 1 \le j \le k\}$. Then

$$\mathscr{G}^{-1}(U) = \{S \in \mathscr{S}'(\hat{R}) \mid \mathscr{G}(S) \in U\} = \{S \in \mathscr{S}'(\hat{R}) \mid \hat{S} \in U\}$$
$$= \{S \in \mathscr{S}'(\hat{R}) \mid |\hat{S}(\phi_j)| \le \varepsilon_j \text{ for } 1 \le j \le k\}$$
$$= \{S \in \mathscr{S}'(\hat{R}) \mid |S(\hat{\phi}_j)| \le \varepsilon_j \text{ for } 1 \le j \le k\}.$$

Since this last set is a weak* neighborhood of zero in $\mathscr{S}'(\hat{R})$ we are done.

Definition 3. For each $T \in \mathscr{S}'(\hat{R})$ we define the Fourier transform of T to be the temperate distribution $\hat{T}$, where $\hat{T}(\phi) = T(\hat{\phi})$ for all $\phi \in \mathscr{S}(R)$ (see Exercises 4.1, problem 4).

Lemma 4. For $f \in L_p(R)$, $1 \le p \le \infty$, let $T_f(\phi) = \int f(x)\phi(x)\, dx$ for every $\phi \in \mathscr{S}(R)$. Then T_f is a temperate distribution in R. Furthermore, if f, g are in $L_p(R)$ and $T_f(\phi) = T_g(\phi)$ for all $\phi \in \mathscr{S}(R)$, then $f = g$.

Proof. If $\phi \in \mathscr{S}(R)$ and $\varepsilon > 0$ are given, we can choose $\sigma > 0$ such that $|(1 + x^2)^k \phi(x)| \le \varepsilon$ for $|x| > \sigma$, and so $|\phi(x)| \le \varepsilon(1 + x^2)^{-k}$ for $|x| > \sigma$. Now $|\int f(x)\phi(x)\, dx| \le \int |f(x)\phi(x)|\, dx$ over $\{x \mid |x| \le \sigma\} + \int |f(x)\phi(x)|\, dx$ over $\{x \mid |x| > \sigma\}$.

The first of these is finite because $\phi(x)$ restricted to $[-\sigma, \sigma]$ is in $L_q(R)$. The second term is $\leq \varepsilon \int |f(x)(1 + x^2)^{-k}| \, dx$ over $\{x \,|\, |x| > \sigma\}$, and this is $\leq \varepsilon \|f\|_p \|(1 + x^2)^{-k}\|_q$, which is finite for k sufficiently large. Thus we have shown that T_f is a linear functional on $\mathscr{S}(R)$ and a similar argument shows that $T_f \in \mathscr{S}'(R)$.

Now let $f, g \in L_p(R)$ and suppose that $T_f = T_g$. Then, in particular, $\int f(x)\phi(x) \, dx = \int g(x)\phi(x) \, dx$ for all $\phi \in C_0^\infty(R)$. Hence, if $p > 1, f = g$ because $C_0^\infty(R)$ is dense in the Banach space $L_q(R)$ (Section 1, Theorem 1). If f, g are in $L_1(R)$ we have $f_\varepsilon(x) = \int f(x - \varepsilon y)\omega(y) \, dy = \varepsilon^{-1} \int f(y)\omega[(x - y)/\varepsilon] \, dy$ and, by hypothesis, this is equal to $\varepsilon^{-1} \int g(y)\omega[(x - y)/\varepsilon] \, dy = g_\varepsilon(x)$; here ω is the function defined in Section 1 just before Lemma 1. Now, as $\varepsilon \to 0, f_\varepsilon$ and g_ε tend to f and g, respectively, for the L_1-norm (Section 1, proof of Theorem 1). Thus $f = g$.

Whenever it is convenient we shall identify the L_p-function f with the temperate distribution T_f. We can, and do, define the Fourier transform of f to be $\hat{T}_f$ (Definition 3 above).

For $f \in L_1(R)$ we defined $\hat{f}(y)$ to be $(1/2\pi)^{1/2} \int f(x)e^{-ixy} \, dx$ and we noted that, in particular, $\hat{f} \in L_\infty(R)$. Let us show that the temperate distribution $\hat{T}_f$ is equal to $T_{\hat{f}}$.

$$\hat{T}_f(\phi) = \int f(x)\hat{\phi}(x) \, dx = \int f(x)\left[(1/2\pi)^{1/2} \int \phi(y)e^{-ixy} \, dy\right] dx$$

(Definition 1). By Fubini's theorem [19, Theorem 19, p. 269] this is equal to

$$\int \phi(y)\left[(1/2\pi)^{1/2} \int f(x)e^{-ixy} \, dx\right] dy = \int \phi(y)\hat{f}(y) \, dy = T_{\hat{f}}(\phi).$$

Thus for functions in $L_1(R)$ our new definition of the Fourier transform is consistent with the one we gave earlier. It follows that the Fourier transform on $L_1(R)$ has an inverse.

Plancherel was the first to show how one could define a Fourier transform on $L_2(R)$. He also showed that the map that takes each $f \in L_2(R)$ to its Fourier transform is an equivalence (Section 3.1, Definition 2) from this space onto itself. The definition of the Fourier transform given by Plancherel is different from the one given above [8, p. 47]; we can however, still get his results.

Theorem 4. If $f \in L_2(R)$ then there is a function $\hat{f} \in L_2(R)$ such that $\hat{T}_f = T_{\hat{f}}$. Furthermore, $\|f\|_2 = \|\hat{f}\|_2$.

Proof. It follows from Theorem 3 that $\|\phi\|_2 = \|\hat{\phi}\|_2$ for each $\phi \in \mathscr{S}(R)$. Thus $|\hat{T}_f(\phi)| \le \|f\|_2 \|\phi\|_2$ for all $\phi \in \mathscr{S}(R)$. Now this inequality holds for all $\phi \in C_0^\infty(R)$ and the latter is norm dense in $L_2(R)$ (Section 1, Theorem 1). Hence $\hat{T}_f$ defines a continuous, linear functional on the Banach space $L_2(R)$. But then there is an $\hat{f} \in L_2(R)$ such that $\|\hat{T}_f\| = \|\hat{f}\|_2$ and $\hat{T}_f(g) = \int g(x)\hat{f}(x)\,dx$ for all $g \in L_2(R)$.

Now work with the temperate distribution $T_{\hat{f}}$ and its Fourier transform. As above, there is a function $\hat{\hat{f}} \in L_2(R)$ such that $\|\hat{T}_{\hat{f}}\| = \|\hat{\hat{f}}\|_2$ and $\hat{T}_{\hat{f}}(g) = \int g(x)\hat{\hat{f}}(x)\,dx$ for all $g \in L_2(R)$. In the first paragraph of this proof we saw that, in particular, $\hat{T}_f = T_{\hat{f}}$ on $C_0^\infty(R)$. Since the latter space is dense in $\mathscr{S}(R)$ (Section 2, Theorem 2) we must have $\hat{T}_f = T_{\hat{f}}$ on $\mathscr{S}(R)$. Thus $\hat{T}_{\hat{f}} = \hat{T}_f$ and so $\int \phi(x)\hat{\hat{f}}(x)\,dx = \int f(x)\hat{\hat{\phi}}(x)\,dx$ for all $\phi \in C_0^\infty(R)$.

If $T \in \mathscr{S}'(R)$, $\phi \in C_0^\infty(R)$ then $\hat{\hat{T}}(\phi) = T(\hat{\hat{\phi}})$ and it is easy to see that $\hat{\hat{\phi}}(x) = \phi(-x)$. Thus

$$\int \phi(x)\hat{\hat{f}}(x)\,dx$$

$$= \int f(x)\hat{\hat{\phi}}(x)\,dx = \int f(x)\phi(-x)\,dx = \int f(-x)\phi(x)\,dx,$$

and so $\int [\hat{\hat{f}}(x) - f(-x)]\phi(x)\,dx = 0$ for all $\phi \in C_0^\infty(R)$. Clearly, this means $\hat{\hat{f}}(x) = f(-x)$ almost everywhere in R; note that $\|\hat{\hat{f}}\|_2 = \|f\|_2$. But $\|\hat{\hat{f}}\|_2 \le \|\hat{f}\|_2 \le \|f\|_2$ by the first part of the proof. We conclude that $\|\hat{f}\|_2 = \|f\|_2$ for all $f \in L_2(R)$.

Corollary. For any $f \in L_2(R)$, $\hat{f}(x)$ is the limit, in the Banach space $L_2(R)$, of $\{(1/2\pi)^{1/2} \int_{-h}^{h} f(y)e^{-ixy}\,dy \,|\, h > 0\}$ as $h \to \infty$.

Proof. For $h > 0$ set $f_h(x) = f(x)$ when $|x| \le h$, and set it equal to zero otherwise. Clearly, $\lim_{h \to \infty} \|f_h - f\|_2 = 0$ and, since $f \to \hat{f}$ is an equivalence on $L_2(R)$, hence $\hat{f}(x)$ is the norm limit of $\hat{f}_h$. In the proof of the theorem we saw that

$$\int \hat{f}_h(x)\phi(x)\,dx$$

$$= \int f_h(x)\hat{\phi}(x)\,dx = \int_{-h}^{h} f(x)\left[(1/2\pi)^{1/2} \int \phi(y)e^{-ixy}\,dy\right]dx$$

for all $\phi \in C_0^\infty(R)$. Since f_h is obviously in $L_1(R)$ we can change the order of integration to get $\int [(1/2\pi)^{1/2} \int_{-h}^{h} f(x)e^{-ixy} dx]\phi(y) dy$. The corollary now follows from the fact that $C_0^{\infty}(R)$ is dense in $L_2(R)$.

Remark. Many of the results of this section are valid when the various function spaces are defined on R^n, $n > 1$. When stated in this more general form they have important applications to the theory of partial differential equations (see [11, 28]).

4. Distributions: Definition and Characterizations

A distribution in Ω (recall that Ω is an open subset of R^n) is a linear functional on $C_0^{\infty}(\Omega)$ that is continuous for a certain locally convex topology on this space. We shall define that topology now.

Let $\mathscr{H}$ be the family of all compact subsets of Ω. For each $K \in \mathscr{H}$ let $\mathscr{D}_K(\Omega) = \{f \in C_0^{\infty}(\Omega) \,|\, \text{supp} f \subset K\}$ and, for each nonnegative integer m, let $p_m(f) = \sup\{|D^s f(x)| \,|\, x \in K, \,|s| \leq m\}$. Each p_m is a seminorm on $\mathscr{D}_K(\Omega)$ (Exercises 1, problem 1), and the family $\{p_m \,|\, m = 0, 1, 2, \ldots\}$ defines a metrizable locally convex topology on this space, which we shall denote by t_K. Also, $\mathscr{D}_K(\Omega)[t_K]$ is a Fréchet space (Exercises 2, problem 1). We can now define a topology on $C_0^{\infty}(\Omega)$ as follows (the terms used were defined in Sections 4.2 and in Exercises 4.2, problem 1b):

From each of the spaces $\mathscr{D}_K(\Omega)$ choose a t_K- neighborhood of zero, U_K, which is absorbing, balanced and convex, and let U be the convex hull of $\bigcup \{U_K \,|\, K \in \mathscr{H}\}$. Let $\mathscr{U}$ be the family of all sets which are constructed in this way.

We are going to show that there is a unique Hausdorff, locally convex topology (we shall call it t^*) on $C_0^{\infty}(\Omega)$ that has $\mathscr{U}$ as a fundamental system of neighborhoods of zero. Before doing that, however, we must mention some of the properties of the sets in this family. For any set S denote the convex hull of S by conv S.

(i)   *If U, V are in $\mathscr{U}$, then $U \cap V$ contains a set in $\mathscr{U}$.*

Proof. We have $U = \text{conv} \bigcup \{U_K \,|\, K \in \mathscr{H}\}$,

$$V = \text{conv} \bigcup \{V_K \,|\, K \in \mathscr{H}\}$$

where each U_K, V_K is an absorbing, balanced, convex neighborhood of zero in $\mathscr{D}_K(\Omega)[t_K]$. Clearly, each of the sets $U_K \cap V_K$ also has these properties and $U \cap V \supset \text{conv} \bigcup \{U_K \cap V_K \,|\, K \in \mathscr{H}\}$.

(ii) $\bigcap \{U \,|\, U \in \mathscr{U}\} = \{0\}$.

Proof. Any $f \in C_0^\infty(\Omega)$ is in some $\mathscr{D}_K(\Omega)$. If f is also in every set in $\mathscr{U}$, then it is in every t_K-neighborhood of zero in this $\mathscr{D}_K(\Omega)$. But since t_K is a Hausdorff topology, $f = 0$.

(iii) *If $U \in \mathscr{U}$ and σ is a positive scalar, then $\sigma U \in \mathscr{U}$.*

Proof. We have $U = \text{conv} \bigcup \{U_K \,|\, K \in \mathscr{H}\}$. So if $y \in U$, then we must have $y = \sum_{i=1}^n \alpha_i y_i$, where each y_i is in some U_K, $\alpha_i \geq 0$, for $i = 1, 2, \ldots, n$, and $\sum_{i=1}^n \alpha_i = 1$ (Exercises 4.2, problem 1b). But then $\sigma y = \sum \alpha_i(\sigma y_i)$, which says σy is in conv $\bigcup \{\sigma U_K \,|\, K \in \mathscr{H}\}$. Since each of the sets σU_K is an absorbing, balanced, convex neighborhood of zero in $\mathscr{D}_K(\Omega)$, this last set is in $\mathscr{U}$. But $\sigma U = \text{conv} \bigcup \{\sigma U_K \,|\, K \in \mathscr{H}\}$; hence σU is in $\mathscr{U}$.

Lemma 1. Let X be a vector space over K and let $\mathscr{U}$ be a family of absorbing, balanced, convex subsets of X that has properties (i), (ii), and (iii) stated above. Then there is a unique Hausdorff, locally convex topology on X that has $\mathscr{U}$ as a fundamental system of neighborhoods of zero.

Proof. For each $U \in \mathscr{U}$ let p_U be the gauge function of U (Section 4.2). By (iii) the family $\{p_U \,|\, U \in \mathscr{U}\}$ satisfies the separation condition (Section 4.1, Lemma 2), and so the topology defined on X by this family is Hausdorff. It is also locally convex. The construction process (Section 4.1) together with properties (i) and (ii) show that $\mathscr{U}$ is a fundamental system of neighborhoods of zero for this topology.

Finally, any two locally convex topologies on X that both have $\mathscr{U}$ as a fundamental system of neighborhoods of zero must coincide (Section 4.3, paragraph after Definition 1).

Remark. Recall the family of sets $\mathscr{U}$ defined on the space $C_0^\infty(\Omega)$ above. There is a unique, Hausdorff, locally convex topology t^* on this space that has $\mathscr{U}$ as a fundamental system of neighborhoods of zero. The space $C_0^\infty(\Omega)$ together with the topology t^* will be called the space of test functions in Ω. The space $C_0^\infty(\Omega)[t^*]$ will be denoted by $\mathscr{D}(\Omega)$.

Definition 1. A continuous, linear functional on $\mathscr{D}(\Omega)$ will be called a distribution in Ω. The set of all distributions in Ω (i.e., the dual of $\mathscr{D}(\Omega)$) will be denoted by $\mathscr{D}'(\Omega)$.

We shall show later on that this terminology is consistent with that of Section 3, Definition 2. Right now we are going to examine t^* in some detail and establish some of its properties. The main result is a very useful necessary and sufficient condition for a linear functional on $C_0^\infty(\Omega)$ to be a distribution in Ω.

Lemma 2. For each $K \in \mathscr{H}$ let j_K be the inclusion map from $\mathscr{D}_K(\Omega)[t_K]$ into $C_0^\infty(\Omega)$. Then t^* is the strongest locally convex topology on $C_0^\infty(\Omega)$ for which each of the maps j_K is continuous.

Proof. If U is any t^*-neighborhood of zero in $\mathscr{D}(\Omega)$, then U contains a set of the form conv $\bigcup \{U_K \mid K \in \mathscr{H}\}$, where each U_K is a t_K-neighborhood of zero in $\mathscr{D}_K(\Omega)$. Since $j_K^{-1}(U) \supset U_K$, it is clear that each j_K is continuous when $C_0^\infty(\Omega)$ has the topology t^*.

Let t be any locally convex topology on $C_0^\infty(\Omega)$ for which each of the maps j_K is continuous, and let U be any absorbing, balanced, convex, t-neighborhood of zero. For each $K \in \mathscr{H}$ define U_K to be $j_K^{-1}(U) = U \cap \mathscr{D}_K(\Omega)$. Since j_K is linear and continuous, each U_K is an absorbing, balanced, convex, t_K-neighborhood of zero in $\mathscr{D}_K(\Omega)$. But clearly $U \supset$ conv $\bigcup \{U_K \mid K \in \mathscr{H}\}$. Since the latter set is a t^*-neighborhood of zero, t is weaker than t^*.

Corollary. Let $G[s]$ be any locally convex space and let g be a linear map from $\mathscr{D}(\Omega)$ into $G[s]$. Then g is continuous iff each of the maps $g \circ j_K$ is continuous.

Proof. Let W be an absorbing, balanced, convex, s-neighborhood of zero in G and assume that each of the maps $g \circ j_K$ is continuous. Then $(g \circ j_K)^{-1}(W) = j_K^{-1}[g^{-1}(W)] = g^{-1}(W) \cap \mathscr{D}_K(\Omega)$ is a t_K-neighborhood of zero in $\mathscr{D}_K(\Omega)$, for every $K \in \mathscr{H}$. Since g is also linear, each of these sets is absorbing, balanced and convex. Clearly, $g^{-1}(W) \supset$ conv $\bigcup \{g^{-1}(W) \cap \mathscr{D}_K(\Omega) \mid K \in \mathscr{H}\}$ and, since the latter set is a t^*-neighborhood of zero, g is continuous.

We can use this corollary to prove that a continuous, linear functional on $\mathscr{S}(R^n)[t_d]$ (Section 2, before Theorem 2) is a distribution in R^n

in the sense of Definition 1. We have maps j_K from $\mathscr{D}_K(R^n)$ into $\mathscr{D}(R^n)$, for each K, and we have an inclusion map I from $\mathscr{D}(R^n)$ into $\mathscr{S}(R^n)$. We shall show that I is continuous by showing that each of the maps $I \circ j_K$ is continuous. If a compact set K, an integer k, and a multi-index s are given, we first let $M = \max\{(1 + |x|^2)^k \mid x \in K\}$ and we note that

$$\max\{|(1 + |x|^2)^k D^s f(x)| \mid x \in K\} \le M \max\{|D^s f(x)| \mid x \in K\}$$

for any $f \in \mathscr{D}_K(R^n)$. It follows from this that I is a continuous, linear map from $\mathscr{D}(R^n)$ into $\mathscr{S}(R^n)[t_d]$. Now the map I^*, defined by $I^*(S) = S \circ I$ for every $S \in \mathscr{S}'(R^n)$, clearly maps $\mathscr{S}'(R^n)$ into $\mathscr{D}'(R^n)$. But, since $I(\mathscr{D}(R^n))$ is t_d-dense in $\mathscr{S}(R^n)$ (Section 2, Theorem 2) this map must be one-to-one.

Before proving our main result of this section we must establish the following lemma. The reader may want to recall the definition of a bounded subset of a locally convex space (Section 4.3, Definition 3).

Lemma 3. Let $X[t]$, $Y[s]$ be two locally convex spaces over the same field, and let T be a linear map from X into Y. Assume that the topology t is metrizable. Then the map T is continuous iff the set $T(B) = \{T(x) \mid x \in B\}$ is s-bounded in Y whenever B is t-bounded in X.

Proof. If T is continuous then, since it is linear, it must map bounded sets to bounded sets (Exercises 4.3, problem 3c). Assume that T satisfies our condition, let U be a balanced, convex, s-neighborhood of zero in Y, and observe that $T^{-1}(U)$ is a balanced, convex subset of X that absorbs all bounded sets (i.e., if $B \subset X$ is a t-bounded set, then for some $\sigma > 0$, $B \subset \sigma[T^{-1}(U)]$).

Let $\{U_n\}$ be a countable fundamental system of t-neighborhoods of zero and assume that $U_n \supset U_{n+1}$ for all n. If $T^{-1}(U)$ is not a t-neighborhood of zero in X, then, for each n, we can choose $x_n \in n^{-1}U_n$ such that $x_n \notin T^{-1}(U)$. Clearly the sequence $\{nx_n\}$ is t-convergent to zero and so it is a t-bounded set. But then $\{nx_n\} \subset \sigma T^{-1}(U)$, which implies that $x_n \in (\sigma/n)T^{-1}(U) \subset T^{-1}(U)$ for all $n \ge \sigma$. This is a contradiction.

Theorem 1. A linear functional T on $C_0^\infty(\Omega)$ is a distribution in Ω iff: To every compact subset K of Ω there corresponds a constant C and an integer k such that $|T(\phi)| \le C \sup\{|D^s \phi(x)| \mid x \in K, |s| \le k\}$ for all $\phi \in \mathscr{D}_K(\Omega)$.

Proof. It is clear (Exercises 1, problem 1) that our condition on T can be restated as follows: To every compact subset K of Ω there corresponds a constant C and an integer k such that $|T(\phi)| \le Cp_k(\phi)$ for all $\phi \in \mathscr{D}_K(\Omega)$.

To prove this theorem we shall use the Corollary to Lemma 2. If $T \in \mathscr{D}'(\Omega)$ then each of the maps $T \circ j_K$ is continuous; i.e., $T \circ j_K$ is a continuous, linear functional on $\mathscr{D}_K(\Omega)[t_K]$. It follows that there is a constant C and an integer k such that $|T \circ j_K(\phi)| \le Cp_k(\phi)$ for all $\phi \in \mathscr{D}_K(\Omega)$. Thus any distribution in Ω satisfies our condition.

Now assume that the linear functional T satisfies our condition. Let B be any t_K-bounded subset of $\mathscr{D}_K(\Omega)$. Since $p_k(B)$ is a bounded set for each k (Exercises 4.3, problem 3a) our condition implies that $T \circ j_K(B)$ is bounded set of complex numbers. It follows from Lemma 3 that $T \circ j_K$ is a continuous, linear functional on $\mathscr{D}_K(\Omega)[t_K]$. But since this is true for every compact set K in Ω the Corollary to Lemma 2 shows that $T \in \mathscr{D}'(\Omega)$.

Theorem 2. A linear functional u on $\mathscr{D}(\Omega)$ is a distribution in Ω iff $\lim u(\phi_j) = 0$ for any sequence $\{\phi_j\}$ of points of $\mathscr{D}(\Omega)$ that has the two following properties:

(i) There is a compact set $K \subset \Omega$ such that supp $\phi_j \subset K$ for all j.

(ii) For any multi-index s the sequence $\{D^s\phi_j(x)\}$ converges to zero uniformly over K.

Proof. Condition (i) says that the sequence $\{\phi_j\}$ is a sequence of points of $\mathscr{D}_K(\Omega)$. Condition (ii) says that $\{\phi_j\}$ tends to zero for the topology of $\mathscr{D}_K(\Omega)$. Now recall the Corollary to Lemma 2, which says that u is continuous on $\mathscr{D}(\Omega)$ iff $u \circ j_K$ is continuous on $\mathscr{D}_K(\Omega)$ for every K; here j_K is the natural inclusion map from $\mathscr{D}_K(\Omega)$ into $\mathscr{D}(\Omega)$. Since each $\mathscr{D}_K(\Omega)$ is a metrizable, locally convex space, $u \circ j_K$ is continuous iff $\lim u \circ j_K(\phi_j) = 0$ whenever $\{\phi_j\}$ is a sequence in $\mathscr{D}_K(\Omega)$ that converges to zero for the topology of this space, i.e., whenever $\{\phi_j\}$ is a sequence in $\mathscr{D}_K(\Omega)$ that satisfies condition (ii). This proves the theorem.

EXERCISES 4

1. Define a sequence of compact subsets of Ω as follows: Let $K_0 = \varnothing$ and, for $k \ge 1$, let $K_k = \{x \in R^n \mid \text{the distance from } x \text{ to}$

$R^n \sim \Omega$ is $\geq 1/k$, and the distance from x to zero is $\leq k$}. Show that K_k is contained in the interior of K_{k+1} for $k = 0, 1, \ldots$ and that $\Omega = \bigcup_{k=0}^{\infty} K_k$.

2. Consider the space $C^{\infty}(\Omega)$. For each compact subset K of Ω and each nonnegative integer m define $p_{K, m}(f)$ to be $\sup\{|D^s f(x)| \, | \, x \in K, \, |s| \leq m\}$ for each f in this space.

   *(a) Show that each $p_{K, m}$ is a seminorm on $C^{\infty}(\Omega)$ and that the family of all such seminorms satisfies the separation condition (Section 4.1, Lemma 2).

   *(b) Let t_0 be the Hausdorff, locally convex topology defined on $C^{\infty}(\Omega)$ by the family $\{p_{K, m} \, | \, K \in \mathcal{H}, \, m = 0, 1, 2, \ldots\}$. Show that t_0 is metrizable. Hint: Use problem 1 above.

   (c) Show that $C^{\infty}(\Omega)[t_0]$ is a Fréchet space.

*3. (a) Let $\{f_n\}$ be a Cauchy sequence in $\mathscr{D}(\Omega)$; i.e., for every continuous seminorm p on this space, $\lim p(f_n - f_m) = 0$ as m, $n \to \infty$. Show that there is a compact subset K of Ω such that $\operatorname{supp} f_n \subset K$ for all n. Hint: Suppose that there is a sequence $\{x_k\}$ in Ω that has no accumulation point, and a subsequence $\{f_k\}$ of $\{f_n\}$ such that $f_k(x_k) \neq 0$ for every k. Let $\{K_k\}$ be an increasing sequence of compact subsets of Ω whose union is Ω, and that satisfies: $K_0 = \varnothing$, $x_k \in K_k \sim K_{k-1}$ for all $k \geq 1$. Define $p(f)$ to be $2 \sum_{k=1}^{\infty} \sup\{f(x)/f_k(x_k) \, | \, x \in K_k \sim K_{k-1}\}$. Finally, show that p is a continuous seminorm on $\mathscr{D}(\Omega)$ but $p(f_n - f_m)$ does not tend to zero as m, $n \to \infty$.

   (b) Use (a) to show that a sequence $\{f_n\}$ of points of $\mathscr{D}(\Omega)$ converges to $f \in \mathscr{D}(\Omega)$ iff: (i) There is a compact subset K of Ω such that $\operatorname{supp} f_n \subset K$ for all n. (ii) For any multi-index s, $\{D^s f_n(x)\}$ converges to $D^s f(x)$ uniformly over K.

4. (a) We refer to the sequence of compact sets $\{K_k\}$ defined in problem 1 above. Let $\mathscr{D}_k(\Omega) = \{f \in C_0^{\infty}(\Omega) \, | \, \operatorname{supp} f \subset K_k\}$ and let t_k be the usual topology on this space. If K is any compact subset of Ω and if $K \subset K_k$ show that the restriction of t_k to $\mathscr{D}_K(\Omega)$ coincides with t_k. In particular, $t_{k+1} \, | \, \mathscr{D}_k(\Omega) = t_k$ for all k.

   (b) For each k we have an inclusion map j_k from $\mathscr{D}_k(\Omega)[t_k]$ into $C_0^{\infty}(\Omega)$. Let t^{**} be the strongest locally convex topology on $C_0^{\infty}(\Omega)$ for which each of these maps is continuous. Show that $t^{**} = t^*$.

   (c) If K is any compact subset of Ω show that the restriction of

$t^* = t^{**}$ to $\mathcal{D}_K(\Omega)$ coincides with t_K. Hint: It suffices to show that $t^*|\mathcal{D}_k(\Omega) = t_k$ for every k. Fix k and let $V_k = \{f \in \mathcal{D}_k(\Omega)\,|\,p_{n(i)}(\phi) \leq \varepsilon_i$ for $1 \leq i \leq n\}$. Then for any p let $V_{k+p} = \{f \in \mathcal{D}_{k+p}(\Omega)\,|\,p_{n(i)}(\phi) \leq \varepsilon_i$ for $1 \leq i \leq n\}$, and let $V = \bigcup_{p=1}^{\infty} V_{k+p}$.   Note that

$$V \cap \mathcal{D}_{k+p}(\Omega) = V_{k+p}, \quad V \cap \mathcal{D}_k(\Omega) = V_k,$$

and, if

$$l < k, V \cap \mathcal{D}_l(\Omega) = [V \cap \mathcal{D}_k(\Omega)] \cap \mathcal{D}_l(\Omega) = V_k \cap \mathcal{D}_l(\Omega).$$

Finally, note that V is a t^*-neighborhood of zero.

(d)   For any compact subset K of Ω show that the space $\mathcal{D}(\Omega)[t_K]$ is a closed, linear subspace of $\mathcal{D}(\Omega)$.

(e)   Show that the topology t^* is not metrizable. Hint: First observe (see problem 3) that if t^* were metrizable then $\mathcal{D}(\Omega)$ would be a Fréchet space. Since $\mathcal{D}(\Omega) = \bigcup_{k=1}^{\infty} \mathcal{D}_k(\Omega)$ and since each $\mathcal{D}_k(\Omega)$ is closed in $\mathcal{D}(\Omega)$ (by (d)), our claim follows from the Baire category theorem.

5.   Distributions: Examples, Properties, and Applications

Here we shall establish the principal properties of distributions and present some examples. We shall use the characterization given in Theorem 1 of the last section as our starting point. Recall: A linear functional T on $C_0^\infty(\Omega)$ is a distribution in Ω iff to every compact subset K of Ω there corresponds a constant C and an integer k such that $|T(\phi)| \leq C \sup\{|D^s\phi(x)|\,|\,x \in K, |s| \leq k\}$ for all $\phi \in \mathcal{D}_K(\Omega)$.

(a) Examples.   (i) A function f in Ω is said to be locally integrable if for every compact subset K of Ω, $\int |f(x)|\,dx < \infty$ (the integration is over the set K). For any such function f define $T_f(\phi) = \int f(x)\phi(x)\,dx$ for all $\phi \in \mathcal{D}(\Omega)$. Then T_f is a distribution in Ω.

Proof.   For any compact subset K of Ω we have

$$|T(\phi)| \leq \int_K |f(x)|\,|\phi(x)|\,dx \leq \left[\int_K |f(x)|\,dx\right] \sup\{|\phi(x)|\,|\,x \in K\}$$

for all $\phi \in \mathscr{D}_K(\Omega)$. So T_f satisfies our criterion for a distribution in Ω where $C = \int_K |f(x)|\ dx$ and $k = 1$.

(ii) We recall that a G-delta in Ω is any subset of Ω that is the intersection of a countable family of open sets, and that a σ-ring is a family of sets that is closed under complements and countable unions. The Baire sets in Ω are the elements of the smallest σ-ring that contains every compact G-delta. If μ is a complex-valued measure on the Baire sets in Ω such that $|\mu|(K) < \infty$ for every compact set K, then the functional T_μ on $\mathscr{D}(\Omega)$, defined by $T_\mu(\phi) = \int \phi(x)\ d\mu(x)$ for every $\phi \in \mathscr{D}(\Omega)$, is a distribution in Ω.

Proof.   For any compact subset K of Ω and any $\phi \in \mathscr{D}_K(\Omega)$ we have

$$| T_\mu(\phi)| \leq \int |\phi(x)|\ d\mu(x) \leq |\mu|(K) \sup\{|\phi(x)|\ |x \in K\}.$$

So T_μ satisfies our criterion for $C = |\mu|(K)$ and $k = 1$.

(iii) For any fixed point $a \in \Omega$ define δ_a on the Baire sets of Ω as follows: $\delta_a(A) = 0$ if $a \notin A$, $\delta_a(A) = 1$ if $a \in A$. By (ii) δ_a defines a distribution in Ω, which we shall denote by T_a. We call T_a the Dirac distribution at the point a.

(b) Multiplication.   For any $T \in \mathscr{D}'(\Omega)$ and any $f \in C^\infty(\Omega)$ we can define a linear functional S on $C_0^\infty(\Omega)$ by letting $S(\phi) = T(f\phi)$ for all ϕ in this space. We claim that any such S is a distribution in Ω.

Proof.   Let K be a compact subset of Ω, let C and k be such that $|T(\phi)| \leq C \sup\{|D^s\phi(x)|\ |x \in K, |s| \leq k\}$ for all $\phi \in \mathscr{D}_K(\Omega)$. Recall that, for any two multi-indices $j = (j_1, j_2, \ldots, j_n)$ and $q = (q_1, \ldots, q_n)$, the symbol $\binom{j}{q}$ means

$$\binom{j_1}{q_1}\binom{j_2}{q_2}\cdots\binom{j_n}{q_n},$$

where

$$\binom{j_i}{q_i} = j_i!/q_i!(j_i - q_i)!.$$

We also recall the Leibnitz formula $D^j(f\phi) = \sum_{q \leq j} \binom{j}{q} D^q f D^{j-q}\phi$, and we let $M = \sup\{|D^s f(x)| \,|\, x \in K, \; |s| \leq k\}$. It follows that

$$|D^j f\phi(x)| \leq M \sum_{q \leq j} \binom{j}{q} |D^{j-q}\phi(x)|$$

$$\leq M \sup\{|D^j\phi(x)| \,|\, x \in K, \; |j| \leq k\} \sum_{q \leq j} \binom{j}{q}.$$

Thus

$$|S(\phi)| \leq C \sup\{|D^j\phi(x)| \,|\, x \in K, \; |j| \leq k\},$$

where $C = M \max\{\sum_{q \leq j} \binom{j}{q}| \,|j| \leq k\}$.

(c) Differentiation. If f, ϕ are in $C_0^\infty((0, 1))$ then, using integration by parts, we find that $\int f'(x)\phi(x)\, dx = -\int f(x)\phi'(x)\, dx$. So if dT/dx denotes the distribution defined by $f'(x)$ (as in (a), example (i)), then we have $dT/dx(\phi) = -T_f(\phi')$ for all ϕ. This observation motivates the following:

Definition 1. Let T be any distribution in Ω. We define $D^s T$ as follows: $D^s T(\phi) = (-1)^{|s|} T(D^s\phi)$ for all $\phi \in \mathscr{D}(\Omega)$.

We must show that $D^s T \in \mathscr{D}'(\Omega)$. However, for any compact subset K of Ω we have C and k such that $|T(\phi)| \leq C \sup\{|D^r\phi(x)| \,|\, x \in K, \; |r| \leq k\}$ for all $\phi \in \mathscr{D}_K(\Omega)$, and clearly $D^s T$ must satisfy a similar inequality.

If $\Omega = R$ and $H(x) = 1$ for $x \geq 0$ and $= 0$ for $x < 0$, then clearly $dT_H/dx(\phi) = -\int H(x)\phi'(x)\, dx = -\int_0^\infty \phi'(x)\, dx = \phi(0) = T_0(\phi)$, where T_0 is the Dirac distribution at zero. So the distributions dT_H/dx and T_0 are equal.

Definition 2. Let U be an open subset of Ω. There is an inclusion map I from $\mathscr{D}(U)$ into $\mathscr{D}(\Omega)$ that is clearly continuous. Hence I^*, where $I^*(T) = T \circ I$ for all $T \in \mathscr{D}'(\Omega)$, maps $\mathscr{D}'(\Omega)$ into $\mathscr{D}'(U)$. If $T \in \mathscr{D}'(\Omega)$, we shall call $T \circ I$ the restriction of T to U. We shall say that S, $T \in \mathscr{D}'(\Omega)$ are equal in U if their restrictions to U are equal.

Theorem 1. Let ω be an open subset of R^n whose closure is compact, and suppose that this closure is contained in Ω. Then for any $T \in \mathscr{D}'(\Omega)$ there is an $f \in L_\infty(\omega)$ and an integer m such that the distributions T and $D_1^m D_2^m D_3^m \cdots D_n^m f$ are equal in ω; here $D_j^m = \partial^m/\partial x_j^m$.

Proof.   Given $T \in \mathcal{D}'(\Omega)$ we seek a function $f \in L_\infty(\omega)$ such that, for any $\phi \in C_0^\infty(\omega)$, we have: (1) $T(\phi) = (D_1^m \cdots D_n^m f)\phi = (-1)^{mn} \int f(x)[D_1^m \cdots D_n^m \phi(x)]\, dx$. If we had such a function f and if we let $C = \|f\|_\infty$, then (1) would yield: (2)

$$|T(\phi)| \le C \int |D_1^m \cdots D_n^m \phi(x)|\, dx$$

for all $\phi \in C_0^\infty(\omega)$.

Now suppose that we have a distribution S in Ω that satisfies (2) for some constant C and some integer m. Clearly

$$\{(-1)^{mn} D_1^m \cdots D_n^m \phi(x) \,|\, \phi \in C_0^\infty(\omega)\}$$

is a linear subspace of $L_1(\omega)$, and inequality (2) says that the map that takes $(-1)^{mn} D_1^m \cdots D_n^m \phi(x)$ to $S(\phi)$ is continuous on this subspace for the L_1-norm. By the Hahn–Banach theorem we can extend this map to a continuous, linear functional on all of $L_1(\omega)$, whose norm is $\le C$. But since $L_\infty(\omega)$ is the dual of $L_1(\omega)$ this means that there is a function $f \in L_\infty(\omega)$ that satisfies (1), and for which $\|f\|_\infty \le C$.

The argument up to this point shows that, to prove the theorem, it suffices to show that any $T \in \mathcal{D}'(\Omega)$ satisfies (2) for some constant C and some integer m. Now the closure of ω, cl ω, is compact, and so there is a constant C and an integer k such that $|T(\phi)| \le C \sup\{|D^s\phi(x)| \,|\, x \in \mathrm{cl}\ \omega, |s| \le k\}$ for all $\phi \in \mathcal{D}_{\mathrm{cl}\ \omega}(\Omega)$. Let $\psi \in C_0^\infty(\omega)$ and, for fixed j, let a_j be the upper bound for $\{|x_j| \,|\, x = (x_1, \ldots, x_n) \in \omega\}$. By the mean value theorem $\sup|\psi(x)| \le a_j \sup|D_j\psi(x)|$ for all such ψ and, by repeated use of this estimate we obtain: (3) $|T(\phi)| \le C' \sup|D_1^k \cdots D_n^k \phi(x)|$. For any $\phi \in C_0^\infty(\omega)$ we can write $\phi(x) = (-1)^n \int D_1 \cdots D_n \phi(x)\, dx$, where the integration is over the set $y < x$, i.e., $y_1 < x_1$, $y_2 < x_2$, $\ldots$, $y_n < x_n$. So (4) $\sup|\phi(x)| \le C' \int \sup|D_1 \cdots D_n \phi(x)|\, dx$. From (3) and (4) we get (2) with $m = k + 1$.

Our next result relates the derivative of a distribution that is defined by a function to the classical derivative of that function. We recall the nonnegative, C^∞-function ω whose support is the unit ball of R^n, which is positive in the interior of that ball, and satisfies $\int \omega(x)\, dx = 1$ (Section 1, before Lemma 1).

Theorem 2.   Let g, f be two continuous functions in Ω and suppose that, as distributions, $D_j g = f$ (here $D_j = \partial/\partial x_j$), i.e., $D_j T_g(\phi) = T_f(\phi)$

for all $\phi \in \mathcal{D}(\Omega)$. Then $D_j g$ exists in the classical sense and is equal to the function $f(x)$.

Proof. We want to prove that $D_j g$ exists at each point $x \in \Omega$ and is equal to $f(x)$. By making suitable use of a function in $C_0^\infty(\Omega)$ that is one on a neighborhood of x we see that we may assume that both g and f have compact support in Ω.

For any $\varepsilon > 0$ let $g_\varepsilon(x) = \varepsilon^{-n} \int g(y)\omega[(x - y)/\varepsilon]\, dy$. We have already seen that $g_\varepsilon(x) \in C_0^\infty(\Omega)$ and that it tends to $g(x)$ uniformly in Ω as ε tends to zero (Section 1, proof of Lemma 1). Now

$$D_j g_\varepsilon(x) = \varepsilon^{-n} \int g(y) D_j \omega\left(\frac{x - y}{\varepsilon}\right) dy = \varepsilon^{-n} \int g(y) \frac{\partial}{\partial x_j} \omega\left(\frac{x - y}{\varepsilon}\right) dy$$

$$= -\varepsilon^{-n} \int g(y) \frac{\partial}{\partial y_j} \omega\left(\frac{x - y}{\varepsilon}\right) dy = \varepsilon^{-n} D_j T_g\left[\omega\left(\frac{x - y}{\varepsilon}\right)\right]$$

$$= \varepsilon^{-n} T_f\left[\omega\left(\frac{x - y}{\varepsilon}\right)\right]$$

by hypothesis. This last term is $\varepsilon^{-n} \int f(y)\omega[(x - y)/\varepsilon]\, dy = f_\varepsilon(x)$. Thus $D_j g_\varepsilon(x)$ tends to $f(x)$ uniformly over Ω as ε tends to zero. It follows that $D_j g(x)$ exists and that it is equal to $f(x)$.

(d) Support. If f is a continuous function in Ω we have defined the support of f, supp f, to be the closure of $\{x \in \Omega \mid f(x) \neq 0\}$. We are going to define the support of a distribution in Ω in such a way that, in particular, the support of T_f coincides with supp f.

We shall say that a distribution T in Ω vanishes on the open subset U of Ω if the restriction of T to U (see Definition 2 above) is zero.

Lemma 1. If a distribution in Ω vanishes on each member of a family of open sets, then it vanishes on the union of this family.

Proof. Let $\{U_v \mid v \in J\}$ be a family of open subsets of Ω, let $T \in \mathcal{D}'(\Omega)$, and assume that T vanishes on U_v for each $v \in J$. We construct an open covering of Ω as follows: Choose $\mu \notin J$, let U be the union of the family $\{U_v \mid v \in J\}$, choose $\phi \in \mathcal{D}(\Omega)$ with supp $\phi \subset U$, and set $U_\mu = \Omega \sim \text{supp } \phi$. Then, setting $I = J \cup \{\mu\}$, we have a covering $\{U_v \mid v \in I\}$ of Ω.

There is a C^∞-partition of unity $\{\alpha_v \mid v \in I\}$ that is subordinate to this covering (Section 1, Theorem 2). Clearly, $\phi = \sum \{\alpha_v \phi \mid v \in I\}$ and so $T(\phi) = \sum \{T(\alpha_v \phi) \mid v \in I\}$. Now if $v \in J$, then $T(\alpha_v \phi) = 0$ by hypothesis. Also, since supp $\alpha_\mu \subset U_\mu = \Omega \sim$ supp ϕ, $\alpha_\mu \phi = 0$ and so $T(\alpha_\mu \phi) = 0$. Thus $T(\phi) = 0$ for every $\phi \in \mathscr{D}(U)$ and this says T vanishes on U.

Definition 3.   For each $T \in \mathscr{D}'(\Omega)$ consider the family of all open subsets W of Ω such that T vanishes on W. Let U be the union of this family. Then we define the support of T, supp T, to be the set $\Omega \sim U$.

Suppose that f is a real-valued, continuous function in Ω and let x be any point at which f is not zero. Then for any neighborhood U of x there is a neighborhood V of x such that $V \subset U$ and f is of constant sign on V. Choose one more open neighborhood (call it W) of x in such a way that, denoting the closure of W by cl W, cl $W \subset V$. There is a function $\phi \in \mathscr{D}(\Omega)$ (Section 1, Corollary to Theorem 2) such that: $0 \le \phi(y) \le 1$ for all $y \in \Omega$; $\phi(z) = 1$ for all $z \in$ cl W; $\phi(y) = 0$ for $y \notin V$. Then $T_f(\phi) = \int f(y)\phi(y)\, dy \ne 0$ and so $x \in$ supp T_f; i.e., we have shown that supp $f \subset$ supp T_f.

Now suppose that $z \notin$ supp f. Choose any neighborhood of z that is disjoint from supp f and notice that $T_f(\phi) = 0$ for all functions ϕ whose support is in this neighborhood of z. Thus $z \notin$ supp T_f. So the support of the function f coincides with the support of the distribution T_f.

We leave it to the reader to compute supp T_f when f is a complex-valued, continuous function in Ω.

The next theorem requires that we recall the topology t_0 defined on $C^\infty(\Omega)$ (Exercises 4, problem 2b).

Theorem 3.   The dual of the locally convex space $C^\infty(\Omega)[t_0]$ is isomorphic to the space of all distributions that have compact support in Ω.

Proof.   Let S be a linear functional on $C^\infty(\Omega)$ that is t_0-continuous on this space. It is convenient to give the proof in steps.

(i) Corresponding to the given linear functional S there is a compact subset K of Ω, a nonnegative integer m, and a constant C such that:

$(*) |S(f)| \le C \sup\{|D^s f(x)| \mid x \in K, |s| \le m\}$ for all $f \in C^\infty(\Omega)$.

Since S is t_0-continuous there is a t_0-neighborhood (call it V) of zero in $C^\infty(\Omega)$ and a constant C such that $|S(f)| \le C$ for all $f \in V$. We may assume, because of the definition of t_0, that there is a compact subset K of Ω and a nonnegative integer m such that $V = \{f \in C^\infty(\Omega) | p_{K,m}(f) \le 1\}$. Condition $(*)$ is equivalent to $|S(f)| \le Cp_{K,m}(f)$ for all $f \in C^\infty(\Omega)$.

If $p_{K,m}(f) = 0$, then the same is true of αf for any scalar α; i.e., $\alpha f \in V$ for all α. But then $|S(\alpha f)| \le C$ for all α implies $S(f) = 0$. Thus $(*)$ is true for any f for which $p_{K,m}(f) = 0$.

Now suppose $p_{K,m}(f) \ne 0$. Then $f/p_{K,m}(f)$ is in V and so $|S(f)| \le Cp_{K,m}(f)$ for any such f.

(ii) Here we shall define a one-to-one, linear map γ from the dual of $C^\infty(\Omega)[t_0]$ into $\mathscr{D}'(\Omega)$.

To the given functional S we have associated a compact set K (condition $(*)$ above). Choose, and fix, $\psi \in C_0^\infty(\Omega)$ such that $\psi(x) = 1$ for all x in some open set containing K (Section 1, Corollary to Theorem 2). For any $f \in C^\infty(\Omega)$, $f = f\psi + (1 - \psi)f$ and so $S(f) = S(f\psi) + S[(1 - \psi)f]$. Since $(1 - \psi)f$ is zero in a neighborhood of K, condition $(*)$ tells us that S vanishes at this function. So we have shown that $S(f) = S(f\psi)$ for all $f \in C^\infty(\Omega)$.

Now regard S as a linear functional on $C_0^\infty(\Omega)$. We recall the inclusion maps j_K from $\mathscr{D}_K(\Omega)$ into $C_0^\infty(\Omega)$ and we also recall that to prove that S is continuous on $\mathscr{D}(\Omega)$ it suffices to show that, for any K, $S \circ j_K(B)$ is a bounded set whenever $B \subset \mathscr{D}_K(\Omega)$ is bounded (Section 4, Corollary to Lemma 2, and Lemma 3). If B is a bounded subset of $\mathscr{D}_K(\Omega)$ then $\{\psi\phi \,|\, \phi \in j_K(B)\}$ is t_0-bounded in $C^\infty(\Omega)$—this follows immediately from the way the seminorms defining t_0 are defined. But since S is t_0-continuous on $C^\infty(\Omega)$, the set $\{S(\psi\phi) \,|\, \phi \in j_K(B)\}$ is bounded. However, $\{S(\psi\phi) \,|\, \phi \in j_K(B)\} = \{S(\phi) \,|\, \phi \in j_K(B)\}$ by the paragraph above. Thus $S \circ j_K$ is continuous for any K and so S defines a distribution $\gamma(S)$ in Ω.

Clearly γ is a linear map from the dual of $C^\infty(\Omega)[t_0]$ into $\mathscr{D}'(\Omega)$. Let us show that γ is one-to-one. If S, T are in this dual space then so is $S - T$. By the first paragraph of the proof of (ii) we can choose $\psi \in C_0^\infty(\Omega)$ such that $(S - T)f = (S - T)\psi f$ for all $f \in C^\infty(\Omega)$. If $\gamma(S) = \gamma(T)$, then $\gamma(S)\psi f = \gamma(T)\psi f$ because $\psi f \in \mathscr{D}(\Omega)$. But then $(S - T)f = 0$ for all $f \in C^\infty(\Omega)$ and so $S = T$.

(iii) We will now show that $\gamma(S)$ has compact support.

Recall the compact set K associated with S (condition $(*)$)) and that $S(f) = S(\psi f)$ for all $f \in C^\infty(\Omega)$, where ψ is any C_0^∞-function that is one on a neighborhood of K. If $\phi \in C_0^\infty(\Omega)$ and supp $\phi \cap K = \varnothing$, then we can choose $\psi \in C_0^\infty(\Omega)$ such that ψ is one on a neighborhood of K, but ψ is zero on supp ϕ. It follows that $\gamma(S)\phi = 0$ and, since this is true of every ϕ with supp $\phi \cap K = \varnothing$, supp $\gamma(S) \subset K$.

(iv) The only thing left to prove is that γ is onto.

Let $S_0 \in \mathscr{D}'(\Omega)$ and let $K = $ supp S_0 be compact. We define a linear functional S on $C^\infty(\Omega)$ as follows: Let ψ be a fixed C^∞-function that is one on a neighborhood of K, and set $S(f) = S_0(\psi f)$ for all $f \in C^\infty(\Omega)$. Let B be any t_0-bounded subset of $C^\infty(\Omega)$ and note that $\{\psi f \mid f \in B\}$ is a bounded subset of $\mathscr{D}(\Omega)$. Since $S_0 \in \mathscr{D}'(\Omega)$ and $\{S(f) \mid f \in B\} = \{S_0(\psi f) \mid f \in B\}$, this set is bounded. It follows that S is t_0-continuous on $C^\infty(\Omega)$ (Exercises 4, problem 2b, and Section 4, Lemma 3). To prove that γ is onto we need only observe that, by the third paragraph of (iii), $\gamma(S) = S_0$.

(e) Convolution.   There does not seem to be any really simple way of treating this topic. However, the usefulness of the convolution operation fully justifies the efforts made to define and study it.

If f, g are C^∞-functions on R and if g has compact support, then the convolution of f and g, $f * g$, is defined as follows:

$$f * g(x) = \int f(x - y)g(y)\, dy = \int f(y)g(x - y)\, dy.$$

Observe that, because of our assumptions on f and g, $f * g$ is well defined, $f * g(x) = g * f(x)$ for all x, and $f * g(x) = T_f[g(x - y)]$, where the distribution T_f is applied to the function $g(x - y)$ of y; here x is fixed. With this as motivation we can define the convolution of a function and a distribution in two important cases.

Definition 4.   Let $\phi \in \mathscr{D}(R^n)$, $u \in \mathscr{D}'(R^n)$. We define the convolution of u and ϕ at x, $u * \phi(x)$, to be $u[\phi(x - y)]$. We will sometimes write $u * \phi(x) = u_y[\phi(x - y)]$, where the notation is meant to emphasize the fact that we are applying the distribution u to the function $\phi(x - y)$ of y; x is fixed.

If $\phi \in C^\infty(R^n)$ and the distribution u has compact support in R^n, then we define $u * \phi(x)$ to be $u_y[\phi(x - y)]$. This has meaning (Section 5, Theorem 3).

From the definition of the support of a distribution it is immediate that $u * \phi(x) = 0$ unless the support of u meets the support of $\phi(x - y)$

(as a function of y); i.e., $u * \phi(x) = 0$ unless there is a $y \in \text{supp } u$ such that $x - y \in \text{supp } \phi$. Thus

$$\text{supp } u * \phi \subset \text{supp } u + \text{supp } \phi.$$

In particular, if both u and ϕ have compact support, then so does $u * \phi$.

Lemma 2. If $\phi \in \mathscr{D}(R^n)$ and $u \in \mathscr{D}'(R^n)$, then $u * \phi \in C^\infty(R^n)$. Also, for any multi-index s, $D^s(u * \phi) = (D^s u) * \phi = u * (D^s \phi)$. The same conclusions hold if the function ϕ is in $C^\infty(R^n)$ and the distribution u has compact support.

Proof. Let $\phi \in \mathscr{D}(R^n)$ and let $\{x^j\}$ be a sequence of points of R^n that converges to the point x. Then the C^∞-functions $\phi_j(y) \equiv \phi(x^j - y)$, $j = 1, 2, 3, \ldots$, converge to the C^∞-function $\phi_0(y) \equiv \phi(x - y)$ in the following manner: (i) There is a compact set K such that supp $\phi_j \subset K$ for all j. (ii) The sequence $\{D^s \phi_j\}$ converges to $D^s \phi_0$ uniformly over K, for any multi-index s. Hence (Section 4, Theorem 2), $\lim u(\phi_j) = u(\phi_0)$ for every $u \in \mathscr{D}'(R^n)$. But this says that $\lim u * \phi(x^j) = u * \phi(x)$; i.e., $u * \phi$ is a continuous function.

In order to prove that $u * \phi \in C^\infty(R^n)$ it suffices to prove the differentiation formulas stated in the hypothesis. First let s be a multi-index such that $|s| = 1$, and let e_k be a unit vector along the positive x_k axis of R^n. Now

$$\{u * \phi(x + he_k) - u * \phi(x)\}h^{-1}$$
$$= u_y[\{\phi(x + he_k - y) - \phi(x - y)\}h^{-1}].$$

If, in this equation, we let h run through a sequence converging to zero, then the quantity in square brackets will converge to $\partial\phi(x - y)/\partial x_k$ in the manner described in (i) and (ii) above. Since h can run through any sequence converging to zero and the conclusion still holds, and since $u \in \mathscr{D}'(R^n)$, the right-hand side of our equation converges to $u * (\partial\phi/\partial x_k)$ as h tends to zero. Hence the left-hand side of our equation also converges as h tends to zero, and clearly its limit is $\partial(u * \phi)/\partial x_k$. The proof can now be completed by induction.

So $u * \phi \in C^\infty$ and $D^s(u * \phi) = u * D^s \phi$ for all s. The fact that $D^s u * \phi = u * D^s \phi$ follows from the definitions.

In case the function ϕ is in $C^\infty(R^n)$ and the distribution u has compact support the proof above goes through almost without change.

Corollary 1. Let $u \in \mathscr{D}'(R^n)$ and, for each $\phi \in \mathscr{D}(R^n)$, let $U(\phi) = u * \phi$. The map U, from $\mathscr{D}(R^n)$ to $C^\infty(R^n)$, has the following property: If $\{\phi_j\}$ is a sequence of points of $\mathscr{D}(R^n)$ that converges, for the topology of $\mathscr{D}(R^n)$, to ϕ_0, then $\lim U(\phi_j) = U(\phi_0)$ for the topology of $C^\infty(R^n)$.

If $\phi \in C^\infty(R^n)$ and the distribution u has compact support, then the map U, $U(\phi) = u * \phi$, from $C^\infty(R^n)$ to $C^\infty(R^n)$ is continuous.

Lemma 3. Let ϕ, ψ be two C^∞-functions on R^n, let u be a distribution in R^n, and assume that any two of these three objects have compact support. Then $(u * \phi) * \psi = u * (\phi * \psi)$.

Proof. Recall that $\phi * \psi(x) = \int \phi(x - y)\psi(y) \, dy$, and since these functions are in $C^\infty(R^n)$ we can approximate the integral by a Riemann sum, say $\varepsilon^n \sum \phi(x - \varepsilon g)\psi(\varepsilon g) \equiv f_\varepsilon(x)$, where $\varepsilon > 0$ is fixed and g runs through all points with integer coordinates. Observe that $f_\varepsilon(x)$ is a C^∞-function of x. Now for any multi-index s, $D^s f_\varepsilon(x) = \varepsilon^n \sum D^s \phi(x - \varepsilon g)\psi(\varepsilon g)$, and this converges to $(D^s \phi * \psi)(x) = [D^s(\phi * \psi)](x)$ uniformly as ε goes to zero. It follows that the sequence $\{f_\varepsilon(x) | \varepsilon = 1/n, n = 1, 2, 3, \ldots\}$ converges to $\phi * \psi$ for the topology of $C^\infty(R^n)$ if only one of the functions ϕ, ψ has compact support, and that it converges to $\phi * \psi$ for the topology of $\mathscr{D}(R^n)$ if both of these functions have compact support. Hence $u * (\phi * \psi) = \lim u * f_\varepsilon$ in either case (Corollary 1 above). Thus $[u * (\phi * \psi)](x) = \lim u * f_\varepsilon(x) = \lim \varepsilon^n \sum (u * \phi)(x - \varepsilon g)\psi(\varepsilon g) = [(u * \phi) * \psi](x)$ for all x.

We have already seen (see (a), example (i)) that $C^\infty(R^n) \subset \mathscr{D}'(R^n)$. Let us use Lemmas 2 and 3 to characterize the closure of this subspace.

Theorem 4. The vector space $C^\infty(R^n)$ is weak* dense in $\mathscr{D}'(R^n)$; i.e., it is dense in $\mathscr{D}'(R^n)$ for the topology $\sigma(\mathscr{D}'(R^n), \mathscr{D}(R^n))$.

Proof. Choose $\omega \in C_0^\infty(R^n)$ with $\omega \geq 0$,

$$\text{supp } \omega = \{x \in R^n \mid |x| \leq 1\}$$

and $\int \omega(x) \, dx = 1$. For $\varepsilon > 0$ let $\omega_\varepsilon(x)$ be $\varepsilon^{-n}\omega(x/\varepsilon)$ and note that $u * \omega_\varepsilon$ is in $C^\infty(R^n)$ for every $u \in \mathscr{D}'(R^n)$ (Lemma 2). Hence, to prove the theorem, all we need do is show that $u * \omega_\varepsilon(\psi)$ converges to $u(\psi)$, as ε tends to zero, for all $\psi \in \mathscr{D}(R^n)$. If $\psi \in \mathscr{D}(R^n)$, then $\check{\psi}(x) \equiv \psi(-x)$ for all x is also in $\mathscr{D}(R^n)$, and $u(\psi) = (u * \check{\psi})(0)$ since $(u * \check{\psi})(0) = u_y[\check{\psi}(0 - y)] = u_y[\psi(y)]$. Thus what we must show is that

$u * \omega_\varepsilon(x) = [(u * \omega_\varepsilon) * \check{\psi}](0)$ tends to $u(\psi)$. But, by Lemma 3, the left-hand side is equal to $[u * (\omega_\varepsilon * \check{\psi})](0)$ and, since $\omega_\varepsilon * \check{\psi}$ converges to $\check{\psi}$, we are done.

For any fixed distribution u we can define a map U as follows: $U(\phi) = u * \phi$ (see Corollary 1 to Lemma 2). It is clear that U is linear and we have already seen that it maps a convergent sequence onto a convergent sequence. This map has one more important property. For any fixed $h \in R^n$ let $\tau_h \phi(x) = \phi(x - h)$ for every $\phi \in C^\infty(R^n)$. Then $\tau_h(u * \phi)(x) = u * \phi(x - h) = u_y[\phi(x - h - y)]$, and

$$[u * (\tau_h \phi)](x) = u_y[\tau_h \phi(x - y)] = u_y[\phi(x - y - h)].$$

Hence $u * \tau_h \phi = \tau_h(u * \phi)$ and it follows that $\tau_h U(\phi) = U(\tau_h \phi)$; i.e., U commutes with translation.

Lemma 4. Let U be a linear map from $\mathscr{D}(R^n)$ into $C^\infty(R^n)$ that commutes with translation and maps convergent sequences onto convergent sequences. Then there is a unique distribution u in R^n such that $U(\phi) = u * \phi$ for all $\phi \in \mathscr{D}(R^n)$.

If U is a continuous, linear map from $C^\infty(R^n)$ into $C^\infty(R^n)$ that commutes with translation, then there is a unique distribution u, with compact support in R^n, such that $U(\phi) = u * \phi$ for all $\phi \in C^\infty(R^n)$.

Proof. For each $\phi \in \mathscr{D}(R^n)$ let $\check{\phi}(x) = \phi(-x)$ for all x. The map that takes $\check{\phi}$ to $U(\phi)(0)$ is a linear functional on $\mathscr{D}(R^n)$. Let us assume, for the moment, that this map is a distribution (call it u) in R^n. Then $U(\phi)(0) = u(\check{\phi})$ and so $u * \phi(0) = u_y[\phi(0 - y)] = u_y[\check{\phi}(y)] = U(\phi)(0)$; i.e., the equation $u * \phi(x) = U(\phi)(x)$ holds when $x = 0$. But $U(\tau_{-h}\phi)(0) = (u * \tau_{-h}\phi)(0) = \tau_{-h}[U(\phi)(0)] = \tau_{-h}[u * \phi(0)]$; i.e.,

$$U(\phi)(h) = u * \phi(h)$$

for all h.

If U is a continuous, linear map from $C^\infty(R^n)$ to $C^\infty(R^n)$, then the map that takes each $\check{\phi}$ ($\phi \in C^\infty(R^n)$) to the number $U(\phi)(0)$ is clearly a continuous, linear functional on $C^\infty(R^n)$. Thus this map is a distribution with compact support in R^n (Section 5, Theorem 3). The remainder of the proof is the same as that given for the case above.

We shall now show that the distribution u defined, as above, by the map U is unique. Suppose that there is another distribution v such that $U(\phi) = v * \phi$ for all $\phi \in \mathscr{D}(R^n)$. Recall the C^∞-function ω that is nonnegative, has for its support the unit ball in R^n, and satisfies $\int \omega(x)\, dx = 1$. For $\varepsilon > 0$ define $\omega_\varepsilon(x)$ to be $\varepsilon^{-n}\omega(x/\varepsilon)$ and recall that $u * \omega_\varepsilon(\phi)$ converges to $u(\phi)$ for every $\phi \in \mathscr{D}(R^n)$ (see the proof of Theorem 4). Thus if $u * \omega_\varepsilon = v * \omega_\varepsilon$ for all $\varepsilon > 0$, then $u = v$.

Let u_1, u_2 be two distributions in R^n. Suppose that u_2 has compact support. Then $u_2 * \phi \in \mathscr{D}(R^n)$ for every $\phi \in \mathscr{D}(R^n)$, and so $u_1 * (u_2 * \phi)$ has meaning and is in $C^\infty(R^n)$ (see the paragraph after Definition 4, and use Lemma 2). The map that takes each $\phi \in \mathscr{D}(R^n)$ to $u_1 * (u_2 * \phi)$ is linear, sequentially continuous (Corollary 1 to Lemma 2), and translation invariant (paragraph before Lemma 4). Hence, by Lemma 4, there is a unique distribution u such that $u_1 * (u_2 * \phi) = u * \phi$ for every $\phi \in \mathscr{D}(R^n)$.

Similarly, if u_1 has compact support (but u_2 need not), then $u_2 * \phi$ is in $C^\infty(R^n)$ and so $u_1 * (u_2 * \phi)$ has meaning and is in $C^\infty(R^n)$ also. Again, there is a distribution u, which is unique, such that $u_1 * (u_2 * \phi) = u * \phi$ for all ϕ.

Definition 5.   Let u_1, u_2 be two distributions in R^n and assume that at least one of them has compact support. Then $u_1 * u_2$ is defined to be the unique distribution u that satisfies $u_1 * (u_2 * \phi) = u * \phi$ for all $\phi \in \mathscr{D}(R^n)$.

Note that, by definition, $(u_1 * u_2) * \phi = u_1 * (u_2 * \phi)$. If u_1, u_2, u_3 are in $\mathscr{D}'(R^n)$ and if all but one of them has compact support, then

$$[(u_1 * u_2) * u_3] * \phi = (u_1 * u_2) * (u_3 * \phi) = u_1 * [u_2 * (u_3 * \phi)]$$

$$= [u_1 * (u_2 * u_3)] * \phi$$

for all $\phi \in \mathscr{D}(R^n)$. So convolution is associative.

Lemma 5.   If u_1, u_2 are in $\mathscr{D}'(R^n)$ and if at least one of these distributions has compact support, then $u_1 * u_2 = u_2 * u_1$.

Proof.   We have:

$$(u_1 * u_2) * (\phi * \psi) = u_1 * [u_2 * (\phi * \psi)]$$

$$= \text{(by Lemma 3) } u_1 * [(u_2 * \phi) * \psi]$$

$$= u_1 * [\psi * (u_2 * \phi)]$$

(convolution of C^∞-functions is commutative) $= (u_1 * \psi) * (u_2 * \phi)$.

Similarly, $(u_2 * u_1) * (\phi * \psi) = (u_1 * \psi) * (u_2 * \phi)$. So

$$(u_1 * u_2) * (\phi * \psi) = (u_2 * u_1) * (\phi * \psi)$$

for all ϕ, ψ in $\mathscr{D}(R^n)$. This says that $[(u_1 * u_2) * \phi] * \psi = [(u_2 * u_1) * \phi] * \psi$ for all ψ and, as we saw in the last paragraph of the proof of Lemma 4, this implies $(u_1 * u_2) * \phi = (u_2 * u_1) * \phi$ for all ϕ. Using this same argument again we obtain $u_1 * u_2 = u_2 * u_1$.

Lemma 6.   Let $u \in \mathscr{D}'(R^n)$ and let T_0 be the Dirac distribution at zero (see (a), example (iii)). Then $D^s u = (D^s T_0) * u$ for any multi-index s. Also, if u_1, $u_2 \in \mathscr{D}'(R^n)$ and at least one of these distributions has compact support, then $D^s(u_1 * u_2) = (D^s u_1) * u_2 = u_1 * (D^s u_2)$ for any multi-index s.

Proof.   Observe that $(u * T_0) * \phi = u * (T_0 * \phi) = u * \phi$ for all $\phi \in \mathscr{D}(R^n)$. Hence $u * T_0 = u$ and so

$$(D^s u) * \phi = u * (D^s \phi) \qquad \text{(Lemma 2)}$$

$$= [u * (D^s \phi)] * T_0 = u * [(D^s \phi) * T_0] = u * [\phi * (D^s T_0)]$$

$$= u * [(D^s T_0) * \phi] = [u * (D^s T_0)] * \phi = [(D^s T_0) * u] * \phi$$

for all ϕ. It follows that $D^s u = (D^s T_0) * u$ and the second part of the lemma follows immediately from this formula.

Let $P(w)$ be a polynomial of degree m in the n variables $w_1, w_2, \ldots, w_n$; so $w = (w_1, w_2, \ldots, w_n) \in R^n$. If we replace each w_j by the operator $D_j = \partial/\partial x_j$, we obtain a partial differential operator, of order m, with constant coefficients. We shall denote this operator by $P(D)$.

Suppose that for a given operator $P(D)$ we have a distribution F in R^n such that $P(D)F$, the operator applied to F, is equal to T_0 (the Dirac distribution at zero). Then, if T is any distribution with compact support in R^n, the equation $P(D)S = T$ can be solved for S since $P(D)(F * T) = [P(D)F] * T$ (by Lemma 6) $= T_0 * T = T$ and so $S = F * T$. The distribution F is called a fundamental solution for the operator $P(D)$.

We shall illustrate these ideas by finding a fundamental solution for the Laplacian in R^n. In order to do this we shall need the next few results.

For any fixed $\varepsilon > 0$ define $\Phi_\varepsilon(\phi)$ to be $\int_{|x| > \varepsilon}[\phi(x)/x]\, dx$ for all $\phi \in \mathscr{D}(R)$. Any such ϕ has compact support and if supp $\phi \subset [-a, a]$, then $|\Phi_\varepsilon(\phi)| \leq \max\{|\phi(x)|\,|\,x \in [-a, a]\}2 \log a/\varepsilon$. Hence $\Phi_\varepsilon \in \mathscr{D}'(R)$.

If $\phi \in \mathscr{D}(R)$ we may write $\phi(x) = \phi(0) + x\psi(x)$, where $\psi(x)$ is a continuous function with $\psi(0) = \phi'(0)$. Choose a so that $\phi(x) = 0$ for $|x| > a$. Then

$$\Phi_\varepsilon(\phi) = \phi(0) \int_{\varepsilon < |x| < a} dx/x + \int_{\varepsilon < |x| < a} \psi(x)\, dx$$

$$= \phi(0)(\log a/\varepsilon - \log a/\varepsilon) + \int_{\varepsilon < |x| < a} \psi(x)\, dx.$$

Thus $\lim_{\varepsilon \to 0^+} \Phi_\varepsilon(\phi) = \int_{-a}^{a} \psi(x)\, dx$ and we may define a linear functional Φ on $\mathscr{D}(R)$ by setting $\Phi(\phi)$ equal to this limit (for each $\phi \in \mathscr{D}(R)$).

Let us show that Φ is a distribution in R. It suffices to show that for every compact subset K of R the map $\Phi \circ j_K$ is continuous on $\mathscr{D}_K(R)$, where, as usual, j_K denotes the natural inclusion map. Choose and fix a compact subset K of R, choose a sequence $\{\varepsilon(n)\}$ of positive numbers such that $\lim_{n \to \infty} \varepsilon(n) = 0$, and let V be a closed, balanced neighborhood of zero in the underlying field C. We can find another closed, balanced neighborhood of zero (call it W) in C such that $W + W \subset V$. Since each $\Phi_{\varepsilon(n)}$ is a distribution in R, $M = \bigcap_{n=1}^{\infty} j_K^{-1} \circ \Phi_{\varepsilon(n)}^{-1}(W)$ is a closed, balanced subset of $\mathscr{D}_K(R)$. Now $\{\Phi_{\varepsilon(n)}(\phi)\,|\,n = 1, 2, \ldots\}$ is a bounded set of complex numbers because $\lim_{n \to \infty} \Phi_{\varepsilon(n)}(\phi)$ exists for each fixed $\phi \in \mathscr{D}(R)$. Hence there is a scalar σ such that $\sigma\Phi_{\varepsilon(n)}(\phi) \in W$ for all n. It follows that $\sigma\phi \in M$ and so M is an absorbing set. Thus $\bigcup_{n=1}^{\infty} nM = \mathscr{D}_K(R)$. But $\mathscr{D}_K(R)$ is a Fréchet space (Exercises 2, problem 1) and so, by the Baire category theorem, M contains an open set. Let U be a neighborhood of zero in $\mathscr{D}_K(R)$ that is contained in $M - M$. We have $\Phi_{\varepsilon(n)} \circ j_K(u) \subset \Phi_{\varepsilon(n)} \circ j_K(M - M) \subset W - W \subset V$ for every n. Hence $\Phi \circ j_K(U) \subset V$ and the proof is complete.

Recall that the Laplacian is the operator $\nabla = \partial^2/\partial x_1^2 + \partial^2/\partial x_2^2 + \cdots + \partial^2/\partial x_n^2$. We shall assume that $n \geq 3$ and we shall show that the following distribution is a fundamental solution for this operator: For each $\phi \in \mathscr{D}(R^n)$ let

$$(1) \qquad F(\phi) = \lim_{\varepsilon \to 0^+} \left[\frac{-\Gamma(n - 2/2)}{4\pi^{n/2}} \int_{|x| > \varepsilon} \frac{\phi(x)}{r^{n-2}}\, dx \right]$$

where $r = (x_1^2 + x_2^2 + \cdots + x_n^2)^{1/2}$ and Γ denotes the gamma function.

From the definition of differentiation ((c), Definition 1) and the fact that $1/r^{n-2}$ is a harmonic function [4, pp. 252–258] we see that

$$(2) \qquad \nabla F(\phi) = F(\nabla \phi) = \frac{-\Gamma(n-2/2)}{4\pi^{n/2}} \lim_{\varepsilon \to 0^+} \int_{r>\varepsilon} \frac{\nabla \phi(x)}{r^{n-2}} \, dx.$$

Choose l so large that the set supp ϕ is contained in the interior of the sphere, centered at zero, of radius l and let Ω be the open set that is bounded by the spheres $r = \varepsilon$ and $r = l$; here ε is a positive number that is less than l. If $\partial \Omega$ denotes the boundary of Ω and $\partial/\partial n$ denotes differentiation in the direction of the outward normal to $\partial \Omega$, then, by one of Green's identities [4, pp. 252–258], we may write

$$(3) \qquad \int_{\Omega} (u \, \nabla v - v \, \nabla u) \, dx = \int_{\partial \Omega} \left(u \, \frac{\partial v}{\partial n} - v \, \frac{\partial u}{\partial n} \right) d\omega,$$

where $d\omega$ is the measure on $\partial \Omega$. In the case of interest here $d\omega = \varepsilon^{n-1} \, d\sigma$ where $d\sigma$ is Lebesgue measure on the unit sphere (call it S) of R^n. So (3) becomes

$$(4) \qquad \int_{r \geq \varepsilon} \frac{\nabla \phi}{r^{n-2}} \, dx = \int_{r \geq \varepsilon} \phi \, \nabla \left(\frac{1}{r^{n-2}} \right) dx + \int_{r=\varepsilon} \phi \, \frac{\partial}{\partial r} \left(\frac{1}{r^{n-2}} \right) \varepsilon^{n-1} \, d\sigma$$

$$- \int_{r=\varepsilon} \frac{1}{r^{n-2}} \frac{\partial \phi}{\partial r} \varepsilon^{n-1} \, d\sigma.$$

The first integral on the right-hand side of (4) is zero because $1/r^{n-2}$ is a harmonic function. The third integral is bounded by some constant times $\varepsilon \int d\sigma$, and so it tends to zero with ε. To treat the second integral we first note that

$$\frac{\partial}{\partial r} \left(\frac{1}{r^{n-2}} \right) = -(n-2)\varepsilon^{1-n} \qquad \text{on} \quad r = \varepsilon.$$

Hence this integral becomes

$$-(n-2) \int_{r=\varepsilon} \phi(x) \, d\sigma = -(n-2)|S|(1/|S|) \int_{r=\varepsilon} \phi(r\sigma) \, d\sigma,$$

where $|S|$ denotes the surface area of S; i.e., $|S| = 2\pi^{n/2}/\Gamma(n/2)$. Thus, if we take the limit, as ε tends to zero, of both sides of (4) we obtain, using (2) and the above remarks, the equation

$$\frac{-4\pi^{n/2}}{\Gamma((n-2)/2)} \nabla F(\phi) = -(n-2)|S|\phi(0) = \frac{-4\pi^{n/2}}{\Gamma((n-2)/2)} T_0(\phi),$$

where, as usual, T_0 denotes the Dirac distribution at zero. It follows from this that F is a fundamental solution for the Laplacian.

If T is any distribution with compact support in R^n, then a solution S of the equation $\nabla S = T$ is $S = F * T$. In particular, if $T = T_f$, where f is a C^∞-function, then

$$S = F * T_f = \frac{-\Gamma((n-2)/2)}{4\pi^{n/2}} \int \frac{f(y)}{|x-y|^{n-2}}\, dy.$$

Solutions to Starred Problems in Chapters 1–4

I have said elsewhere that the first four chapters of this text are introductory. If that is the case, then my practice of leaving some results for the student and then referring to these results later, a practice I consider quite reasonable in the latter part of the book, can be legitimately criticized. An introductory chapter should contain all but the most routine of details. For this reason I have asked one of my graduate students, Andrea Blum, to write up her solutions to the starred problems in Chapters 1–4.

EXERCISES 1.1

Problem 1. By the triangle inequality $\|x - y + y\| \leq \|x - y\| + \|y\|$. Hence $\big|\|x\| - \|y\|\big| \leq \|x - y\|$ for all x, y in E. If $\{x_n\}$ is a sequence of points of $(E, \|\cdot\|)$ that converges to $x_0 \in E$, $\|x_n - x_0\|$

tends to zero as n tends to infinity. Thus $|\,\|x_n\| - \|x_0\|\,|$ tends to zero also.

Problem 2a. If we assume that u is continuous at zero and if $\{x_n\}$ is a sequence of points of E that converges to x_0, then $\lim u(x_n - x_0) = 0$ because $x_n - x_0\}$ converges to zero. By the linearity of u, this is equivalent to $\lim u(x_n) = u(x_0)$; i.e., u is continuous at x_0.

Problem 2b. Assume that there is a constant M such that $\|u(x)\| \le M$ for all x in the unit ball of E. If $\{x_n\}$ is any sequence of nonzero points of E that is convergent to zero, then $\|u(x_n/\|x_n\|)\| \le M$ for all n, or $\|u(x_n)\| \le M\|x_n\|$ for all n. Since $\lim\|x_n\| = 0$ we see that $\lim u(x_n) = 0$ in E. Thus u is continuous at zero and hence, by problem 2a, continuous on E.

Now assume that u is continuous on E. Suppose that for each positive integer n there is an $x_n \in E$ such that $\|x_n\| \le 1$ and $\|u(x_n)\| \ge n$. Choose such an x_n for $n = 1, 2, 3, \ldots$. Then the sequence $\{x_n/\|u(x_n)\|\}$ converges to zero in $(E, \|\cdot\|)$. Since $u(x_n/\|u(x_n)\|)$ has norm one for each n, we have contradicted the continuity of u at zero. Hence there must be a number M such that $\|u(x)\| \le M$ for all $x \in E$ with $\|x\| \le 1$.

Problem 2c. If there is a constant M such that $\|u(x)\| \le M\|x\|$ for all $x \in E$, then clearly u is continuous at zero and hence on E. Conversely, if u is continuous on E then there is an M such that $\|u(x)\| \le M$ for all x in the unit ball of E (by problem 2b). If x is any element of E that is not in this ball, then $x/\|x\|$ is in the unit ball. Hence $\|u(x/\|x\|)\| \le M$ or $\|u(x)\| \le M\|x\|$ for all $x \in E$.

EXERCISES 1.3

Problem 2c. It is clear that c_0 is a linear subspace of l_∞. Suppose that $\{x^n \mid n = 1, 2, \ldots\}$ is a sequence of points of c_0 that converges to $x \in l_\infty$ for the l_∞-norm. We want to show that $x \in c_0$. Here $x^n = \{x_k^n \mid k = 1, 2, \ldots\}$ for each n and $x = \{x_k \mid k = 1, 2, \ldots\}$. For any $\varepsilon > 0$ we can find an integer n_0 such that $\|x^n - x\|_\infty < \varepsilon$ for all $n \ge n_0$; i.e., $\sup\{|x_k^n - x_k| \mid k = 1, 2, \ldots\} < \varepsilon$ for $n \ge n_0$. It follows that, for each fixed k, $\lim x_k^n = x_k$. Also, if l is any integer, $|x_l| \le |x_l - x_l^n| + |x_l^n| < \varepsilon + |x_l^n|$ for all $n \ge n_0$. Fix $n \ge n_0$. Then since $\{x_k^n\} \in c_0$ for this

n, $\lim |x_l^n| = 0$ (as l goes to infinity). Hence $\limsup |x_l| \le \varepsilon$, but since $\varepsilon > 0$ is arbitrary, $\lim x_l = 0$. So we have shown that $x \in c_0$ and also that c_0 is closed in $(l_\infty, \|\cdot\|_\infty)$.

We will now show that $\{e_i \mid i = 1, 2, \ldots\}$ is a Schauder basis for $(c_0, \|\cdot\|_\infty)$. Let $x = \{x_n \mid n = 1, 2, \ldots\}$ be any element of c_0. Then $\|x - \sum_{n=1}^l x_n e_n\|_\infty = \sup\{|x_{l+1}|, |x_{l+2}|, \ldots\}$. Since $x \in c_0$ this supremum tends to zero as l tends to infinity. So for any $x \in c_0$ we have found numbers $\{x_n\}$ such that $\lim \|x - \sum_{n=1}^l x_n e_n\|_\infty = 0$. The uniqueness of these numbers is clear.

Problem 3. The sequence $\{e_n\}$ is not a Hamel basis for l_1 because the element $\{1/n^2\} \in l_1$ cannot be written as the linear combination of a finite number of these vectors.

Given $x = \{x_n\} \in l_1$, $\|x - \sum_{n=1}^l x_n e_n\|_1 = \sum_{k=l+1}^\infty |x_k|$. Since $\sum_{k=1}^\infty |x_k|$ is finite, the latter sum tends to zero as l tends to infinity. So we have found numbers $\{x_n\}$ such that $\lim \|x - \sum_{n=1}^l x_n e_n\|_1 = 0$. It is clear that these numbers are unique.

Problem 4a. Let N be a proper, linear subspace of X. Suppose that N is the null space of some nonzero element f of $X^\#$. Then there is a point $x_0 \in X$ such that $f(x_0) = t$, $t \ne 0$. Clearly $f(x_0/t) = 1$ and so we may assume that $f(x_0) = 1$ to begin with. It is obvious that $x_0 \notin N$ but that, for any $x \in X$, $x - f(x)x_0$ is in N. Hence for each x in X we may write $x = [x - f(x)x_0] + f(x)x_0$, which shows that N has codimension one in X.

Now suppose that N has codimension one in X. Then we can find $x_0 \in X$ such that for any $x \in X$ there is a $y \in N$ and a scalar λ satisfying $x = y + \lambda x_0$. Since N is proper the vector x_0 cannot be in N. Define a map f from X into the underlying field as follows: $f(y) = 0$ for all $y \in N$, $f(x_0) = 1$, and f is linear. If $x \in X$ can be written $x = y_1 + \lambda_1 x_0$ and $x = y_2 + \lambda_2 x_0$, where y_1, y_2 are in N and λ_1, λ_2 are scalars, then $y_2 - y_1 = (\lambda_1 - \lambda_2)x_0$. But since we have already noted that $x_0 \notin N$, we must conclude that $\lambda_1 - \lambda_2 = 0$ and hence that $y_1 = y_2$. So f is well defined. Clearly the null space of f contains N. If x is in the null space of f, then $x = y + \lambda x_0$ where $\lambda = 0$ and $y \in N$; i.e., $x \in N$.

Problem 4b. Let $N(\theta)$, $N(\phi)$ denote the null spaces of θ and ϕ, respectively. We are assuming that $N(\theta) \subset N(\phi)$. The first thing we shall show is that $N(\theta) = N(\phi)$. Suppose that there is a point x' that is in $N(\phi)$ but not in $N(\theta)$. Since $N(\theta)$ has codimension one in X we can find $x_0 \in X$ such that for every x there is a $y \in N(\theta)$ and a scalar λ for which $x = y + \lambda x_0$. Then $x' = y' + \lambda' x_0$ for some $y' \in N(\theta)$ and some scalar λ'. Since $\theta(x') \neq 0$, $\lambda' \neq 0$. But $0 = \phi(x') = \lambda' \phi(x_0)$ and hence $\phi(x_0)$ must be zero. It follows that $\phi \equiv 0$ and this is a contradiction. So $N(\theta) = N(\phi)$.

If x is any element of X then $x = y + \lambda x_0$ where $y \in N(\theta) = N(\phi)$ and λ is a scalar. So $\theta(x) = \lambda \theta(x_0)$, or $\lambda = \theta(x)/\theta(x_0)$. Then $\phi(x) = \lambda \phi(x_0) = [\theta(x)/\theta(x_0)]\phi(x_0) = [\phi(x_0)/\theta(x_0)]\theta(x)$. Since $\phi(x_0)/\theta(x_0)$ is a scalar we are done.

EXERCISES 2.2

Problem 6b. Let X be a vector space, let $\{\phi_1, \phi_2, \ldots, \phi_p\}$ be a finite, linearly independent subset of $X^\#$, and let ϕ_0 be an element of $X^\#$ whose null space contains $\bigcap_{j=1}^{p} N(\phi_j)$; here $N(\phi_j)$ is the null space of ϕ_j for each $j = 1, 2, \ldots, p$. We claim that the set $\{\phi_0, \phi_1, \ldots, \phi_p\}$ is linearly dependent.

The first thing we shall establish is that the space $\bigcap_{j=1}^{p} N(\phi_j)$ has codimension p in X. We shall prove this by induction on p. The case $p = 1$ has already been considered (Exercises 1.3, problem 4a, solved above). Assume that the result is true when $p = k$ and consider a linearly independent set $\{\phi_1, \phi_2, \ldots, \phi_{k+1}\}$ containing $k + 1$ elements. Let $N_k = \bigcap_{j=1}^{k} N(\phi_j)$ and note that, by the inductive hypothesis, N_k has codimension k in X. Consider the restriction of ϕ_{k+1} to N_k; call it ϕ'_{k+1}. We distinguish two cases: (i) $N(\phi_{k+1}) \supset N_k$ so that $\phi'_{k+1} = 0$; (ii) $N(\phi'_{k+1}) = N(\phi_{k+1}) \cap N_k$ is a proper, linear subspace of N_k.

Case (i). It is clear that in this case $\phi_{k+1} \in (X/N_k)^\#$. Now X/N_k has dimension k and, since k is finite, $(X/N_k)^\#$ also has dimension k. But $\{\phi_1, \phi_2, \ldots, \phi_k\}$ is a linearly independent subset of $(X/N_k)^\#$ that contains k elements. Thus ϕ_{k+1} is a linear combination of these elements and we have contradicted the linear independence of the set $\{\phi_1, \phi_2, \ldots, \phi_k, \phi_{k+1}\}$. So case (i) cannot arise.

Case (ii). We have already observed that N_k has codimension k in X. This means that there is a k-dimensional subspace X_k of X such that $X = N_k \oplus X_k$. Hence given any point $x \in X$ there are unique elements $y \in N_k$, $z \in X_k$ such that $x = y + z$. Now by Exercises 1.3, problem 4a (solved above), the space $N(\phi'_{k+1})$ has codimension one in N_k. So there is an element $x' \in N_k$, $x' \notin N(\phi'_{k+1})$, such that $N_k = N(\phi'_{k+1}) \oplus$

$\text{lin}\{x'\}$. The element y, mentioned above, can be written, uniquely, as $y = y' + \lambda x'$ for some $y' \in N(\phi'_{k+1})$ and some scalar λ. Thus $x = y' + (\lambda x' + z)$ where $y' \in N(\phi'_{k+1})$ and $\lambda x' + z$ is in the linear span of $\{x', X_k\}$. Since $x' \notin X_k$, this linear span has dimension $k + 1$, and hence $N(\phi'_{k+1}) = N(\phi_{k+1}) \cap N_k$ has codimension $k + 1$ in X.

We now return to problem 6b. Let $N_p = \bigcap_{j=1}^{p} N(\phi_j)$ and note that N_p has codimension p in X. Since p is finite both X/N_p and $(X/N_p)^\#$ have dimension p. But the set $\{\phi_1, \phi_2, \ldots, \phi_p\}$ is contained in $(X/N_p)^\#$, is linearly independent, and contains p elements. Thus this set spans $(X/N_p)^\#$. Since $\phi_0 \in (X/N_p)^\#$ it is a linear combination of $\phi_1, \ldots, \phi_p$, and so $\{\phi_0, \phi_1, \ldots, \phi_p\}$ is a linearly dependent set.

EXERCISES 2.4

Problem 3. Let B be a Banach space and let G be a closed, linear subspace of B that has a complement in B. In order to show that any two complements of G are topologically isomorphic it suffices to prove that any complement, say H, of G is topologically isomorphic to the Banach space B/G. There is a continuous projection operator P from B onto H with null space G. Define a map π from B/G onto H as follows: If $\dot{x} \in B/G$ let $\pi(\dot{x}) = P(x)$ where x is any element of $\dot{x}$. Clearly π is linear. Also, since P maps B onto H, π is onto. If $\pi(\dot{x}) = \pi(\dot{y})$, then $P(x) = P(y)$ for all $x \in \dot{x}$ and all $y \in \dot{y}$. It follows that $x - y \in G$, which says $\dot{x} = \dot{y}$. So π is one-to-one. Once we have shown that π is continuous then, by the open-mapping theorem, we will have shown that π is a topological isomorphism. Clearly, $\|\pi(\dot{x})\| \leq \|P\| \, \|x\|$ for every $x \in \dot{x}$. Thus $\|\pi(\dot{x})\| \leq \|P\| \inf\{\|x\| \, | \, x \in \dot{x}\}$. But, by definition, $\inf\{\|x\| \, | \, x \in \dot{x}\} = \|\dot{x}\|$. Hence $\|\pi(\dot{x})\| \leq \|P\| \, \|\dot{x}\|$.

EXERCISES 2.5

Problem 2. We have two closed, linear subspaces G and H of a Banach space $(B, \|\cdot\|)$. Assume that $G \cap H = \{0\}$ and that there is a scalar α such that $\|g\| \leq \alpha \|g + h\|$ for all g in G and all $h \in H$. We want to prove that $G + H$ is closed. Let $\{g_n + h_n\}$ be a sequence of points of $G + H$ that converges to $z \in B$. Then given $\varepsilon > 0$ we have

$$\|g_n - g_m\| \leq \alpha \|(g_n - g_m) + (h_n - h_m)\|$$
$$= \alpha \|(g_n + h_n) - (g_m + h_m)\| \leq \alpha \varepsilon$$

for m, n sufficiently large. It follows that $\{g_n\}$ is a Cauchy sequence of points of G, and hence that $\lim g_n$, which we shall call g_0, is a point of G. Since $\lim\{g_n + h_n\} = z$ and $\lim g_n = g_0$ we see that $\lim h_n$ must exist. Call this h_0 and note that $h_0 \in H$. But then $g_0 + h_0 = z$, which says $G + H$ is closed.

EXERCISES 3.4

Problem 1a. Let $(B_1, \|\cdot\|_1)$ and $(B_2, \|\cdot\|_2)$ be two completions of $(E, \|\cdot\|)$. We may regard E as a dense linear subspace of $(B_1, \|\cdot\|_1)$ and also of $(B_2, \|\cdot\|_2)$, and $\|\cdot\|_1 = \|\cdot\| = \|\cdot\|_2$ on E. If $x \in B_1$, $x \notin E$, then there is a sequence of points of E (call it $\{x_n\}$) that converges to x for $\|\cdot\|_1$. Define a map ϕ from B_1 onto B_2 as follows: For $x \in E$ let $\phi(x) = x$, for $x \in B_1$ but not in E let $\phi(x) = \lim \phi(x_n)$ for $\|\cdot\|_2$.

Since $\{x_n\}$ is a Cauchy sequence in $(E, \|\cdot\|)$ and $\phi(x_n) = x_n$ for all n, it is clear that $\lim \phi(x_n)$ exists in $(B_2, \|\cdot\|_2)$. Furthermore, if $\{x_n\}, \{z_n\}$ are two sequences of points of E and if $\lim x_n = \lim z_n$ in $(B_1, \|\cdot\|_1)$, then $\{x_n - z_n\}$ converges to zero for $(E, \|\cdot\|)$. Thus $\{\phi(x_n) - \phi(z_n)\}$ also converges to zero for $(E, \|\cdot\|)$ and this says that $\lim \phi(x_n) = \lim \phi(z_n)$ in $(B_2, \|\cdot\|_2)$. Thus ϕ is well defined.

It is easy to see that ϕ is linear. Suppose $y \in B_2$, $y \notin E$, is given. There is a sequence $\{y_n\}$ of points of E that converges to y for $\|\cdot\|_2$. But then $\{y_n\}$ is Cauchy for $\|\cdot\|$, and hence $\{y_n\}$ converges to some element $x \in B_1$ for $\|\cdot\|_1$. Clearly, $\phi(x) = y$ and so the map ϕ is onto. Suppose $\phi(x) = \phi(z)$ for two elements x, z in B_1. Choose sequences $\{x_n\}, \{z_n\}$ of points of E such that $\lim x_n = x$, $\lim z_n = z$ for $\|\cdot\|_1$. Then $\{\phi(x_n) - \phi(z_n)\}$ converges to zero for $\|\cdot\|_2$. It follows that $\{x_n - z_n\}$ converges to zero for $\|\cdot\|_1$. Hence $x = z$ and ϕ is one-to-one.

The only thing left to prove is that $\|\phi(x)\|_2 = \|x\|_1$ for all $x \in B_1$. Given $x \in B_1$ choose $\{x_n\} \subset E$ such that $\lim x_n = x$. Then since $\|\phi(x_n)\|_2 = \|x_n\| = \|x_n\|_1$ for all n, and since $\|\phi(x)\|_2 = \lim \|\phi(x_n)\|_2$ while $\|x\|_1 = \lim \|x_n\|_1$, we see that ϕ is an equivalence.

Problem 1b. Let $(B, \|\cdot\|)$ be the completion of $(E, \|\cdot\|)$. We shall show that the Banach spaces B' and E' are equivalent. Let $f \in E'$. For any two elements x, y of E we have $|f(x) - f(y)| \leq \|f\|\,\|x - y\|$. It follows that f is uniformly continuous on $(E, \|\cdot\|)$. Now E is dense in $(B, \|\cdot\|)$, and hence f has a unique, uniformly continuous extension to all of B. Call this extension $\hat{f}$. We recall that $\hat{f}(x) = f(x)$ for all $x \in E$, and that if $x \in B$ but is not in E then $\hat{f}(x) = \lim f(x_n)$, where $\{x_n\}$ is any sequence of points of E that converges to x. It is clear from this that $\hat{f}$ must be linear whenever f is linear, and so if $f \in E'$, $\hat{f} \in B'$. Define u from E' into B' as follows: For each $f \in E'$ let $u(f) = \hat{f}$. It is trivial to show that u is linear, one-to-one, and onto. Since $\|f\| = \sup\{|f(x)| \mid x \in E, \|x\| \leq 1\}$ and $\|\hat{f}\| = \sup\{|\hat{f}(y)| \mid y \in B, \|y\| \leq 1\}$, in order to show that $\|u(f)\| = \|f\|$ for all $f \in E'$ it suffices to show that the unit ball of E is dense in the unit ball of B.

Let $\mathscr{B}$ be the unit ball of B, $\mathscr{B}(E)$ the unit ball of E. We must show that each $y \in \mathscr{B}$ is the limit of a sequence of points of $\mathscr{B}(E)$. Let $y \in \mathscr{B}$ and let $\{x_n\}$ be any sequence of points of E that converges to y. There are two cases: (i) $\|y\| < 1$; (ii) $\|y\| = 1$. In case (i) we can choose N so that $\|x_n - y\| < 1 - \|y\|$ for all $n \geq N$. Then $\|x_n\| \leq \|x_n - y\| + \|y\| < 1$ for all $n \geq N$, and so $\{x_n \mid n \geq N\}$ is a sequence in $\mathscr{B}(E)$ that converges to y.

In case (ii) let $S = \{n \mid \|x_n\| > 1\}$ and let $T = \{n \mid \|x_n\| \leq 1\}$. If T is an infinite set, then $\{x_m \mid m \in T\}$ is a sequence of points of $\mathscr{B}(E)$ that converges to y; so we may assume that T is a finite set, and hence S is an infinite set. Let $y_m = x_m \|x_m\|^{-1}$ for every $m \in S$. Then $\{y_m\} \subset \mathscr{B}(E)$, $\lim y_m = \lim x_m \lim \|x_m\|^{-1} = \lim x_m \cdot 1 = y$.

EXERCISES 4.1

Problem 1a. Let $U \in \mathscr{V}$. We may assume that there is a finite number of seminorms $p_1, p_2, \ldots, p_n$ in the given family, and a finite number of positive numbers $\varepsilon_1, \varepsilon_2, \ldots, \varepsilon_n$ such that $U = \{x \in X \mid p_i(x) < \varepsilon_i \text{ for } 1 \leq i \leq n\}$. Define $V \in \mathscr{V}$ as follows: $V = \{x \in X \mid p_i(x) < \varepsilon_i/2 \text{ for } 1 \leq i \leq n\}$. We want to show that $V + V \subset U$. Let $x + y \in V + V$ and note that $p_i(x + y) \leq p_i(x) + p_i(y) < \varepsilon_i$ for $1 \leq i \leq n$.

Problem 1b.   Let U be as in problem 1a. Let $N = \{\alpha \in K \mid |\alpha| \leq 1\}$ and let $V = U$. Then $\alpha V \subset U$ for all $\alpha \in N$.

Problem 1c.   Clearly the sequence $\{x_n\}$ is t-convergent to x_0 iff $\{x_n - x_0\}$ is t-convergent to zero. We shall show that $\{x_n\}$ is t-convergent to zero iff $\lim p_\gamma(x_n) = 0$ for every p_γ. First assume that $\{x_n\}$ is t-convergent to zero. For each k let $V_k = \{x \mid p_\gamma(x) < 1/k\}$. Then we must have $x_n \in V_k$ for all n sufficiently large. Thus $p_\gamma(x_n) < 1/k$ for all n. Now since k is arbitrary, $\lim p_\gamma(x_n) = 0$.

Now suppose that $\lim p_\gamma(x_n) = 0$ for every p_γ. Given any t-neighborhood U of zero we may assume that $U = \{x \mid p_j(x) < \varepsilon_j$ for $1 \leq j \leq k\}$. We can find n_1 such that $p_1(x_n) < \varepsilon_i$ for $n \geq n_1$. Similarly, we can find n_2 such that $p_2(x_n) < \varepsilon_2$ for $n \geq n_2$, etc. Choose n_0 greater than any of the numbers $n_1, n_2, \ldots, n_k$. Then $p_j(x_n) < \varepsilon_j$ for all $n \geq n_0$ and for $j = 1, 2, \ldots, k$; i.e., $x_n \in U$ for all $n \geq n_0$.

i.e. span $S \neq E$

Problem 3a.   Suppose that S is a total subset of E and that lin S (the linear span of S) is not dense in E. Then the closure of lin S, call it cl(lin S), is a proper, closed, linear subspace of E. By the Hahn–Banach theorem there is an element $f \in E'$ that is not zero on all of E but is zero on all of cl(lin S). But then f vanishes on S but not on all of E, contradicting the fact that S is total.

Now suppose that lin S is dense in E. If $f \in E'$ vanishes on S, then it must vanish on lin S; this is a consequence of the linearity of f. But then f vanishes on a dense subset of E. Since a continuous, linear functional on E is obviously uniformly continuous on E, this is impossible.

Problem 3b.   If E is separable, then any countable, dense subset of E is a countable, total subset of E. Suppose that E contains a countable, total subset S. By problem 3a the space lin S is dense in E. Thus given $\varepsilon > 0$ and $x \in E$ there is an element $y \in \text{lin } S$ such that $\|x - y\| < \varepsilon$. Now $y = \sum_{j=1}^{n} \alpha_j s_j$ where $s_1, \ldots, s_n$ are in S and $\alpha_1, \ldots, \alpha_n$ are scalars–let us say real scalars. Let $M = \max\{\|s_j\| \mid j = 1, 2, \ldots, n\}$ and for each j let q_j be a rational number such that $|\alpha_j - q_j| < \varepsilon/nM$. Then

$$\|x - \sum q_j s_j\|$$
$$\leq \|x - \sum \alpha_j s_j\| + \|\sum \alpha_j s_j - \sum q_j s_j\|$$
$$< \varepsilon + \sum |\alpha_j - q_j| \, \|s_j\| < 2\varepsilon.$$

So the set of all linear combinations, with rational coefficients, of elements of S is dense in E. Since this set is countable, E is separable.

Problem 4a.   Let u be a linear map from $(E, \|\cdot\|)$ into $(F, \|\|\cdot\|\|)$.

Define u^* from F' into E' as follows: For each $\phi \in F'$, $u^*(\phi)$ is the element of E' defined by $u^*(\phi)x = \phi[u(x)]$ for all $x \in E$. We call u^* the adjoint of u.

For ϕ_1, ϕ_2 in F' and any scalars α, β we have

$$u^*(\alpha\beta_1 + \beta\phi_2)x = (\alpha\phi_1 + \beta\phi_2)[u(x)] = \alpha\phi_1[u(x)] + \beta\phi_2[u(x)]$$
$$= \alpha[u^*(\phi_1)(x)] + \beta[u^*(\phi_2)(x)]$$

for all $x \in E$. Hence $u^*(\alpha\phi_1 + \beta\phi_2) = \alpha u^*(\phi_1) + \beta u^*(\phi_2)$, and we have shown that u^* is linear.

Suppose that u is continuous and give F', E' their norm topologies. Since u is continuous there is a constant M such that $|\,\|u(x)\|\,| \leq M\|x\|$ for all $x \in E$. Now $\sup\{\|u^*(\phi)\|\,|\,\phi \in F', \|\phi\| \leq 1\}$, and

$$\|u^*(\phi)\| = \sup\{|u^*(\phi)x|\,|\,x \in E, \|x\| \leq 1\}$$
$$= \sup\{|\phi[u(x)]|\,|\,x \in E, \|x\| \leq 1\} \leq \|\phi\|\,\|u\|.$$

Thus $\sup\{\|u^*(\phi)\|\,|\,\phi \in F', \|\phi\| \leq 1\} \leq \|u\| \leq M$. It follows that u^* is continuous, and also that $\|u^*\| \leq \|u\|$.

We will now show that we actually have $\|u^*\| = \|u\|$. For any $\phi \in F'$ and any $x \in E$ we have $|\phi[u(x)]| \leq \|\phi\|\,|\,\|u(x)\|\,|$. If x is fixed, then we can find $\phi \in F'$ such that $\|\phi\| = 1$ and $\phi[u(x)] = |\,\|u(x)\|\,|$ (by the Hahn–Banach theorem). It follows that $\sup\{|\phi[u(x)]|\,|\,\|\phi\| \leq 1\} = |\,\|u(x)\|\,|$; here x is any element of E. Using the definition of u^* we may write $\sup\{|u^*(\phi)x|\,|\,\|\phi\| \leq 1\} = |\,\|u(x)\|\,|$. But

$$\sup\{|u^*(\phi)x|\,|\,\|\phi\| \leq 1\} \leq \sup\{\|u^*\|\,\|\phi\|\,\|x\|\,|\,\|\phi\| \leq 1\} \leq \|u^*\|\,\|x\|.$$

So for every $x \in E$ we have $|\,\|u(x)\|\,| \leq \|u^*\|\,\|x\|$. It follows that $\|u\| \leq \|u^*\|$.

Finally, we want to show that u^* is continuous when E' and F' have their weak* topologies. Let U be a weak* neighborhood of zero in E'. We may assume that there is a finite set $x_1, x_2, \ldots, x_n$ of points of E and positive numbers $\varepsilon_1, \varepsilon_2, \ldots, \varepsilon_n$ such that $U = \{f \in E'\,|\,|f(x_i)| \leq \varepsilon_i$ for $1 \leq i \leq n\}$. Then

$$V = \{g \in F'\,|\,u^*(g) \in U\} = \{g \in F'\,|\,|u^*(g)x_i| \leq \varepsilon_i \text{ for } 1 \leq i \leq n\}$$
$$= \{g \in F'\,|\,|g[u(x_i)]| \leq \varepsilon_i \text{ for } 1 \leq i < n\}$$

is a weak* neighborhood of zero in F'.

Problem 4b. We will now assume that u is an equivalence and we shall prove that u^* is an equivalence. By problem 4a we know that u^* is linear, and that it is continuous when E' and F' have their norm topologies. If $u^*(\phi) = 0$ for some $\phi \in F'$, then $u^*(\phi)x = 0$ for all $x \in E$.

Thus $\phi[u(x)] = 0$ for all $x \in E$. But u is onto and so ϕ must be the zero functional; i.e., u^* is one-to-one.

Let us now show that u^* is onto. Let $f \in E', f \neq 0$, and let $N(f)$ be the null space of f. We know that $N(f)$ has codimension one in E and, since u is an equivalence, clearly $u[N(f)] = \{u(x) \mid x \in N(f)\}$ is a closed, linear subspace of codimension one in F. It follows that there is a $g \in F'$ such that $N(g) = u[N(f)]$. Consider $u^*(g) \in E'$. Clearly, $u^*(g)x = 0$ means $g[u(x)] = 0$, and so $N[u^*(g)]$ contains $N(f)$. Hence $u^*(g) = \lambda f$ for some nonzero scalar λ. But then $u^*(\lambda^{-1}g) = f$ and we have shown that u^* is onto.

Finally, we shall show that u^* is norm preserving. If $g \in F'$, then $|u^*(g)x| = |g[u(x)]| \leq \|g\| \|u(x)\| = \|g\| \|x\|$ for all $x \in E$. So $\|u^*(g)\| \leq \|g\|$. Also, $\|g\| = \sup\{|g(y)| \mid y \in F,$

$$\|y\| \leq 1\} = \sup\{|g[u(x)]| \mid x \in E,$$

$\|x\| \leq 1\}$ because u is an equivalence. Hence $\|g\| \leq \|g \circ u\| = \|u^*(g)\|$.

Problem 5. Let $\{f_n\}$ be any bounded sequence in E' and let $\{x_i\}$ be any countable, dense subset of E. We have $|f_n(x_1)| \leq \|f_n\| \|x_1\| \leq \|x_1\|$, where we assume 1 is a bound for $\{\|f_n\|\}$, for all n. Thus there is a subsequence $\{f_n^1\}$ of $\{f_n\}$ such that $\{f_n^1(x_1)\}$ converges. Now $\{f_n^1(x_2)\}$ is a bounded sequence of complex numbers, and so there is a sub-sequence $\{f_n^2\}$ of $\{f_n^1\}$ such that $\{f_n^2(x_2)\}$ is convergent. Clearly $\{f_n^2(x_1)\}$ is also convergent.

After $\{f_n^1\}, \{f_n^2\}, \ldots, \{f_n^k\}$ have been chosen, observe that $\{f_n^k(x_{k+1})\}$ is a bounded sequence of complex numbers, and hence that there is a subsequence $\{f_n^{k+1}\}$ of $\{f_n^k\}$ such that $\{f_n^{k+1}(x_j)\}$ is convergent for $j = 1, 2, \ldots, k+1$.

Now let $g_1 = f_1^1, g_2 = f_2^2, g_3 = f_3^3, \ldots, g_k = f_k^k, \ldots$. It is clear that, after the kth term, $\{g_n\}$ is a subsequence of $\{f_n^k\}$, and that this is true for $k = 1, 2, \ldots$. Hence $\{g_n(x_k)\}$ converges for each $k = 1, 2, \ldots$. Now $\{g_n\}$ is contained in the unit ball of E'. This ball is $\sigma(E', E)$-compact and so $\{g_n\}$ must have a $\sigma(E', E)$-adherent point $g_0 \in E'$. Since $\{g_n(x_k)\}$ converges its limit must be $g_0(x_k)$ for $k = 1, 2, \ldots$.

Suppose that $x \in E$ and $\varepsilon > 0$ are given. First choose x_j such that $\|x - x_j\| < \varepsilon/3$. Then

$$|g_0(x) - g_n(x)| \leq |g_0(x) - g_0(x_j)|$$
$$+ |g_0(x_j) - g_n(x_j)| + |g_n(x_j) - g_n(x)|$$
$$\leq \|g_0\| \|x - x_j\| + |g_0(x_j) - g_n(x_j)|$$
$$+ \|g_n\| \|x_j - x\|.$$

The first and last terms are $\leq \varepsilon/3$. The middle term can be made $\leq \varepsilon/3$ by taking n sufficiently large.

EXERCISES 4.2

Problem 1b. Let S be a subset of X and let $\mathscr{F}$ be the family of all convex subsets of X that contain S; $\mathscr{F}$ is not empty because $X \in \mathscr{F}$. By definition the convex hull of S (we shall denote it by conv (S)) is $\bigcap \{C \,|\, C \in \mathscr{F}\}$. Define G to be the set $\{\sum_{j=1}^{n} \alpha_j x_j \,|\, n$ is a positive integer, $x_1, \ldots, x_n$ are in S, $\alpha_1, \ldots, \alpha_n$ are nonnegative scalars with $\sum_{j=1}^{n} \alpha_j = 1\}$. Clearly $S \subset G$.

We shall show that G is convex. Let $x, y \in G$ and let γ be a scalar between zero and one. We must show that $\gamma x + (1 - \gamma)y$ is in G. Clearly, $x = \sum_{j=1}^{n} \alpha_j x_j$ and $y = \sum_{i=1}^{m} \beta_i y_i$. Since all the x_j's and all the y_i's are in S all we have to do, in order to prove that $\gamma x + (1 - \gamma)y$ is in G, is to show that $\sum_{j=1}^{n} \gamma \alpha_j + \sum_{i=1}^{m} (1 - \gamma)\beta_i = 1$. But this is easy because the first term is just γ and the second is $1 - \gamma$.

It follows that conv$(S) \subset G$. To prove the reverse inclusion it suffices to show that for any $C \in \mathscr{F}$, $G \subset C$. Let $\sum_{j=1}^{n} \alpha_j x_j \in G$ and let $C \in \mathscr{F}$. Clearly $x_1, \ldots, x_n$ are in C because they are in S. So all we have to do is show that, for any finite subset $x_1, \ldots, x_n$ of a convex set C and any nonnegative scalars $\alpha_1, \ldots, \alpha_n$ whose sum is one, we have $\sum_{j=1}^{n} \alpha_j x_j \in C$. We shall prove this by induction on n.

If $n = 1$ the result is trivial, and when $n = 2$ it is true by the definition of convexity. Assume that it is true when $n = k - 1$ and consider

$$(*) \qquad \sum_{j=1}^{k} \alpha_j x_j = \sum_{j=1}^{k-1} \alpha_j x_j + \alpha_k x_k = \sum_{j=1}^{k-1} \alpha_j x_j + \left(1 - \sum_{j=1}^{k-1} \alpha_j\right) x_k$$

because $\sum_{j=1}^{k} \alpha_j = 1$. Let $\beta = \sum_{j=1}^{k-1} \alpha_j$ and clearly we may assume $\beta \neq 0$. Since $\sum_{j=1}^{k-1} (\alpha_j/\beta) = 1$ we can write the right-hand side of $(*)$ as follows: $\sum_{j=1}^{k-1} \alpha_j (\sum_{j=1}^{k-1} \alpha_j/\beta) x_j + (1 - \sum_{j=1}^{k-1} \alpha_j) x_k$. By our induction hypothesis the term $\sum_{j=1}^{k-1} (\alpha_j/\beta) x_j$ is in C; denote this sum by y. So we now can write the right-hand side of $(*)$ as: $\beta y + (1 - \beta) x_k$, and this is clearly in C.

EXERCISES 4.3

Problem 1a. Let $X[t]$ be a locally convex space, let $x_0 \in X$, and define $g(x)$ to be $x + x_0$ for all $x \in X$. It is clear that g is a one-to-one mapping from X onto itself. Since $(x, y) \to x + y$ is a continuous function from $X \times X$ onto X, g is continuous. To prove that g is a homeomorphism we need only show that the continuous map h, defined by $h(x) = x - x_0$ for all $x \in X$, is the inverse of g. But $h[g(x)] = h[x + x_0] = (x + x_0) - x_0 = x$ and

$$g[h(x)] = g[x - x_0] = (x - x_0) + x_0 = x.$$

Problem 1b. Assume that f is continuous at zero and let x be any point of X, V any neighborhood of $f(x)$. Then $V - f(x)$ is a neighborhood of zero and so there is a neighborhood of U of zero such that $f(U) \subset V - f(x)$. Now $U + x$ is a neighborhood of x and $f(U + x) \subset f(U) + f(x) \subset V - f(x) + f(x) = V$.

We will now show that f is continuous at zero iff it is bounded on some t-neighborhood U of zero. If we assume that f is continuous at zero then $\{x \in X \mid |f(x)| < 1\}$ is a t-neighborhood of zero on which f is bounded. Conversely, let us assume that $|f(U)| < M$ for some t-neighborhood U of zero. Then given $\varepsilon > 0$ we have $|f(x)| < \varepsilon$ for all x in the t-neighborhood $(\varepsilon/M)U$ of zero. Hence f is continuous at zero.

Problem 1c. We need only show that (iii) implies (i). So we assume that $f \neq 0$ is a linear functional on X, that the null space of f, $N(f)$, is not dense in $X[t]$, and that f is not continuous at zero. Our assumptions imply the existence of a nonempty open set G such that $G \cap N(f) = \varnothing$. If $x \in G$ we can find a balanced t-neighborhood U of zero such that $x + U \subset G$. Now by problem 1b above, $f(U)$ is not bounded. In fact, $f(U)$ is the entire field K. So there is a point x' in U such that $f(x') = -f(x)$. Clearly, $x + x'$ is in $N(f)$. But $x + x'$ is also in $x + U \subset G$, and this is impossible because $G \cap N(f) = \varnothing$.

Problem 3a. We have $X[t]$, where $t = t(\{p_\gamma\})$ and $\{p_\gamma \mid \gamma \in \Gamma\}$ is a family of seminorms on X. We want to show that $B \subset X$ is t-bounded iff $\sup\{p_\gamma(x) \mid x \in B\}$ is finite for every γ. Suppose that B is t-bounded. Then, for any p_γ, the set $U(p_\gamma) = \{x \in X \mid p_\gamma(x) \leq 1\}$ is a t-neighborhood of zero. Hence there is a scalar λ such that $B \subset \lambda U(p_\gamma)$. But this says $p_\gamma(x) \leq \lambda$ for all $x \in B$.

Now suppose that B satisfies our condition and let U be a t-neighborhood of zero. We may assume that there is a finite subfamily $p_1, p_2, \ldots, p_n$ in $\{p_y\}$ and positive numbers $\varepsilon_1, \ldots, \varepsilon_n$ such that $U = \{x \in X \mid p_j(x) \le \varepsilon_j \text{ for } 1 \le j \le n\}$. Now $\sup p_j(B) \le M_j$ for $j = 1, 2, \ldots, n$ by assumption. Hence $p_j[(\varepsilon_j M_j^{-1})B] \le \varepsilon_j$ for $j = 1, 2, \ldots, n$. Let $\varepsilon = \min\{\varepsilon_j\}$, $M = \max\{M_j\}$. Then $(M\varepsilon^{-1})U$ contains B.

Problem 3c. Let ϕ be a continuous, linear map from $X[t]$ into $Y[s]$. Assume that B is a t-bounded subset of X. We shall show that the set $\phi(B)$ is an s-bounded subset of Y. Let U be any s-neighborhood of zero in Y. Since ϕ is linear $(\phi(0) = 0)$ and continuous, $\phi^{-1}(U)$ is a neighborhood of zero in $X[t]$. Now B is t-bounded and so there is a scalar λ such that $B \subset \lambda\phi^{-1}(U)$. But then $\phi(B) \subset \lambda U$ and we are done.

Reflexive Banach Spaces

We have characterized the reflexive spaces as those Banach spaces in which the unit ball is compact for the weak topology (Section 4.4, Theorem 2). Equivalently, they are the Banach spaces in which the unit ball is countably compact for the weak topology (Section 5.2, Theorem 3 and Remark 2). There are many, many other characterizations of this class of spaces (see, for example, [30, pp. 69–72]). Of all of these the most striking is that due to James [33]. He proved that a Banach space B is reflexive iff every element of B' attains its supremum over the unit ball of B. A continuous, linear functional that attains its supremum over the unit ball of a Banach space is said to attain its norm. For any Banach space B the set of all elements of B' that attain their norms is dense in B' for the norm topology [29].

Many writers have treated more general classes of Banach spaces that contain, and are in some way similar to, the reflexive spaces. For example, it is clear that every reflexive Banach space is a dual space (Section 5.1). A thorough, and very beautiful, discussion of dual spaces can be found in the classic paper of Dixmier [32]. Equivalent forms of some of Dixmier's results were obtained (apparently independently) by Ruston [34]. Ruston's approach is illuminating and his proofs are different from those of Dixmier. We shall mention only one result. A Banach space B is a dual space iff there is a closed, linear subspace Q in B' that has the two following properties: (a) If $\mathscr{B}'$ denotes the unit ball of B' then $Q \cap B'$ is $\sigma(B', B)$-dense in $\mathscr{B}'$. (b) The unit ball of B is compact for the topology $\sigma(B, Q)$ [32, Théorème 17].

A closed, linear subspace of B' that has property (a) is said to be a subspace of characteristic one in B'. Such a subspace need not have property (b) (see [32, 31, Theorem 4]). We proved a theorem about the completion of a locally convex space that enabled us to connect the work of James with that of Dixmier to obtain the following result: A separable Banach space B is a dual space iff B' contains a subspace of characteristic one each element of which attains its norm [31]. The analogous statement for nonseparable Banach spaces is false.

For more results about dual spaces as well as other generalizations of reflexivity we refer the reader to the literature.

References

Chapter 1–7

1. Apostol, T. M. "Mathematical Analysis," 2nd Ed. Addison–Wesley, Reading, Massachusetts (1974).
2. Banach, S. "Theory of Linear Operations." Monografje Matematyczne Tom I, Warsaw (1932).
3. Brown, A. L., and Page, A. "Elements of Functional Analysis." Van Nostrand–Reinhold, Princeton, New Jersey (1970).
4. Courant, R., and Hilbert, D. "Methods of Mathematical Physics," Vol. II. Wiley (Interscience), New York (1962).
5. Eberlein, W. F. Weak compactness in Banach spaces, I, *Proc. Nat. Acad. Sci. U.S.A.* **33** (1947), 51–53.
6. Enflo, P. A counterexample to the approximation problem in Banach spaces, *Acta Math.* **130** (1973), 309–317.
7. Friedman, A. "Partial Differential Equations." Holt, New York (1969).
8. Goldberg, R. R. "Fourier Transforms." Cambridge Univ. Press, London and New York (1961).
9. Hewitt, E., and Ross, K. A. "Abstract Harmonic Analysis," Vol. I. Springer-Verlag, Berlin and New York (1963).
10. Hewitt, E., and Stromberg, K. "Real and Abstract Analysis." Springer-Verlag, Berlin and New York (1969).

161

11. Hörmander, L. "Linear Partial Differential Operators." Springer-Verlag, Berlin and New York (1969).
12. James, R. C. A non-reflexive Banach space isometric with its second conjugate space, *Proc. Nat. Acad. Sci. U.S.A.* **37** (1951), 174–177.
13. Jeffery, R. L. "Trigonometric Series." Univ. of Toronto Press, Toronto, Canada (1956).
14. Katznelson, Y. "An Introduction to Harmonic Analysis." Wiley, New York (1968).
15. Kelley, J. L. "General Topology." Van Nostrand-Reinhold, Princeton, New Jersey (1955).
16. Köthe, G. "Topological Vector Spaces I." Springer-Verlag, Berlin and New York (1969).
17. Krein, M., and Milman, D. On extreme points of regularly convex sets, *Studia Math.* **9** (1940), 133–138.
18. Lindenstrass, J., and Tzafriri, L. On the complemented subspaces problem, *Israel J. Math.* **9** (1971), 263–269.
19. Marti, J. T. "Introduction to the Theory of Bases." Springer-Verlag, Berlin and New York (1969).
20. Phillips, R. S. On linear transformations, *Trans. Amer. Math. Soc.* **48** (1940), 516–541.
21. Royden, H. L. "Real Analysis," 2nd Ed. Macmillan, New York (1968).
22. Singer, I. "Bases in Banach Spaces I." Springer-Verlag, Berlin and New York (1970).
23. Sobczyk, A. Projections of the space (m) on its subspace (c_0), *Bull. Amer. Math. Soc.* **47** (1941), 938–947.
24. Tricomi, F. G. "Integral Equations." Wiley (Interscience), New York (1957).
25. Whitely, R. Projecting m onto c_0, *Amer. Math. Monthly* **73** (1966), 285–286.
26. Whitely, R. An elementary proof of the Eberlein–Smulian theorem, *Math. Ann.* **172** (1967), 116–118.
27. Widder, D. V. "The Laplace Transform." Princeton Univ. Press, Princeton, New Jersey (1946).
28. Yosida, K. "Functional Analysis," 2nd Ed. Springer-Verlag, Berlin and New York (1968).

Appendix B

29. Bishop, E., and Phelps, R. R. The support functionals of a convex set, Convexity, *Proc. Symp. Pure Math.* **7** (1963), 27–35.
30. Day, M. M. "Normed Linear Spaces," 3rd Ed. Springer-Verlag, Berlin and New York (1973).
31. DeVito, C. L. A completeness theorem for locally convex spaces and some applications, *Math. Ann.* **177** (1968), 221–229.
32. Dixmier, J. Sur un théorème de Banach, *Duke Math. J.* **15** (1948), 1057–1071.
33. James, R. C. Characterizations of reflexivity, *Studia Math.* **23** (1964), 205–216.
34. Ruston, A. F. Conjugate Banach spaces, *Proc. Cambridge Phil. Soc.* **53** (1957), 576–580.

Index

DATE DUE